Lecture Notes in Intelligent Transportation and Infrastructure

Series Editor

Janusz Kacprzyk, Systems Research Institute, Polish Academy of Sciences, Warszawa, Poland

The series "Lecture Notes in Intelligent Transportation and Infrastructure" (LNITI) publishes new developments and advances in the various areas of intelligent transportation and infrastructure. The intent is to cover the theory, applications, and perspectives on the state-of-the-art and future developments relevant to topics such as intelligent transportation systems, smart mobility, urban logistics, smart grids, critical infrastructure, smart architecture, smart citizens, intelligent governance, smart architecture and construction design, as well as green and sustainable urban structures. The series contains monographs, conference proceedings, edited volumes, lecture notes and textbooks. Of particular value to both the contributors and the readership are the short publication timeframe and the world-wide distribution, which enable wide and rapid dissemination of high-quality research output.

More information about this series at http://www.springer.com/series/15991

Andras Varhelyi · Vidas Žuraulis ·
Olegas Prentkovskis
Editors

Vision Zero for Sustainable Road Safety in Baltic Sea Region

Proceedings of the International Conference "Vision Zero for Sustainable Road Safety in Baltic Sea Region", 5–6 December 2018, Vilnius, Lithuania

Editors
Andras Varhelyi
Department of Technology and Society
Lund University
Lund, Sweden

Vidas Žuraulis
Department of Automobile Engineering
Vilnius Gediminas Technical University
Vilnius, Lithuania

Olegas Prentkovskis
Department of Mobile Machinery
and Railway Transport
Vilnius Gediminas Technical University
Vilnius, Lithuania

ISSN 2523-3440 ISSN 2523-3459 (electronic)
Lecture Notes in Intelligent Transportation and Infrastructure
ISBN 978-3-030-22374-8 ISBN 978-3-030-22375-5 (eBook)
https://doi.org/10.1007/978-3-030-22375-5

This Springer imprint is published by the registered company Springer Nature Switzerland AG
The registered company address is: Gewerbestrasse 11, 6330 Cham, Switzerland

Preface

In this volume of “Lecture Notes in Intelligent Transportation and Infrastructure”, we are pleased to present the proceedings of the International Conference “Vision Zero for Sustainable Road Safety in Baltic Sea Region”, which took place in Vilnius, Lithuania, from 5 to 6 December 2018. This event is organized by Vilnius Gediminas Technical University (VGTU) and Lithuanian-Swedish Academy. The mission of the conference is to promote a more comprehensive approach supporting new ideas, technologies, theories, its development and applications, as well as work in progress and activities on all theoretical and practical issues arising in safety of road traffic, transportation and sustainable mobility. An outstanding scientific, society policy and practice-oriented conference was organized to facilitate the spread and exchange of knowledge, skills and attitudes between experts, researchers and companies, aiming to facilitate the implementation of Vision Zero in the Baltic Sea Region.

Road safety is a major social issue. In Europe, more than 26,000 people die in traffic every year. For every traffic-related death on Europe’s roads, there is an estimation of four permanently disabling injuries, eight serious injuries and fifty minor injuries. In addition to human suffering, this is causing the society extensive costs for health care, rehabilitation and sick leaves.

The aim of the conference is to highlight road safety in the Baltic Sea Region by cooperation between Sweden, Norway and Lithuania. Shared experience of road safety from Sweden and Norway for the last 20 years had among the lowest traffic-related death rate in the Europe.

At the conference, the current and future plans of road safety and its impact on the society were presented by representatives from ministries and authorities responsible for traffic and road safety in Sweden, Norway and Lithuania. Latest research from universities, research institutes, science parks and the latest state-of-the-art technology development from the vehicle industry was discussed.

The aim of the proceedings is to highlight the issues of road safety in topics of: policy and implementation, technology and innovation, humans and safety, urban planning and vulnerable road users. As the mobility of the future needs to be sustainable, safe and efficient, the road infrastructure safety management, automated transport and digital systems are also analyzed.

The programme committee of the International Conference "Vision Zero for Sustainable Road Safety in Baltic Sea Region", the organizers and the editors of the proceedings would like to acknowledge the participation of all reviewers who helped to refine contents of this volume and evaluated conference submissions. Our thanks go to all members of programme committee:

- **Peje Michaelsson**, Lithuanian-Swedish Academy
- As. Prof. **Vidas Žuraulis**, Vilnius Gediminas Technical University, Lithuania
- Prof. **Andras Vorhelyi**, Lund University, Sweden
- Prof. **Olegas Prentkovskis**, Vilnius Gediminas Technical University, Lithuania
- As. Prof. **Raimundas Junevičius**, Vilnius Gediminas Technical University, Lithuania
- Prof. **Saugirdas Pukalskas**, Vilnius Gediminas Technical University, Lithuania
- Dr. **Vidmantas Pumputis**, Lithuanian Ministry of Transport and Communication, Lithuania
- **Viktoras Lapinas**, Lithuanian Road Administration, Lithuania
- Dr. **Matts-Åke Belin**, Swedish Transport Administration, Sweden
- Dr. **Sonja Forward**, Swedish National Road and Transport Research Institute (VTI), Sweden
- **Giedrė Ivinskienė**, Lithuanian Transport Safety Administration, Lithuania
- **Mindaugas Katkus**, Lithuanian Road and Transport Research Institute, Lithuania
- **Ove Pettersson**, Ugglasand AB, Sweden
- **Eireen Therese Bjørk**, Statens Vegvesen, Norway
- Dr. **Rune Elvik**, Norwegian Centre for Transport Research, Norway
- As. Prof. **Alona Rauckiene-Michaelsson**, Lithuanian-Swedish Academy and Klaipeda University, Lithuania
- **Lina Lazerenko**, Forensic Science Centre of Lithuania, Lithuania
- Prof. **Valentin Ivanov**, Technische Universität Ilmenau, Germany
- Prof. **Carlo Giacomo Prato**, University of Queensland, Australia
- As. Prof. **Andrus Aavik**, Tallinn University of Technology, Estonia
- Prof. **Dago Antov**, Tallinn University of Technology, Estonia
- Prof. **Mareks Mezītis**, Riga Technical University, Latvia

Thanking all the authors who have chosen "Vision Zero for Sustainable Road Safety in Baltic Sea Region" as the publication platform for their research, we would like to express our hope that their papers will help in further developments in design and analysis of complex systems, offering a valuable and timely resource for scientists, researchers, practitioners and students who work in these areas.

Andras Varhelyi
Vidas Žuraulis
Olegas Prentkovskis

Contents

Analysis of Drivers' Eye Movements to Observe Horizontal Road Markings Ahead of Intersections

Anton Pashkevich[1(✉)], Tomasz E. Burghardt[2], Ksenia Shubenkova[3], and Irina Makarova[3]

[1] Politechnika Krakowska, ul. Warszawska 24, 31-155 Kraków, Poland
apashkevich@pk.edu.pl
[2] M. Swarovski GmbH, Industriestraße 10, 3300 Amstetten, Austria
[3] Kazan Federal University, Syuyumbike Prosp. 10a, Naberezhnye Chelny 423812, Russia

Abstract. Analysis of eye movements of young drivers travelling in an urban environment on two-lane dual carriageway roads, as they approach intersections, was performed with accent on responses to pavement markings. Preview distances of 35–70 m and 70–180 m were evaluated. In each case, the visual region was divided into three sections: centre – view ahead, horizontal line markings, and other areas. Horizontal road signalisation was observed seldom, by only 18–24% test participants, and mostly at the level of gazes, which suggests that the markings are used as a confirmation of other visual cues needed to properly position the vehicle. Measured were dissimilarities between different intersections: gazes from larger distance prevailed in cases of broad roadways, while the opposite was measured with smaller crossings. Fixations on horizontal markings were recorded rarely, but if they occurred, they were rather long: 0.18 s, as compared to 0.24–0.30 s for other visual regions. The gazes were recorded less frequently amongst experienced drivers, but if they occurred their number was not influenced by the drivers' experience or sex.

Keywords: Horizontal road markings · Eye tracking · Driver eye movements

1 Background

1.1 Horizontal Road Markings and Road Safety

Safety of road users belongs to one of the key contemporary issues. World Health Organisation estimated that over 3,400 people die every day as a result of road accidents [1]. The same report estimates that financial expenses associated with the accidents reach approximately 3% of World's Gross Domestic Product. Thus, increase in road safety appears to be of critical importance and could improve standard of life for all people [2]. To road features that can meaningfully influence safety belong horizontal road markings, which play important role in supporting drivers' in their travel path [3]. Their installation and maintenance was calculated to be on average sixty times less expensive than the cost of chaos and accidents on the roads in their absence [4].

A. Varhelyi et al. (Eds.): VISZERO 2018, LNITI, pp. 1–10, 2020.
https://doi.org/10.1007/978-3-030-22375-5_1

1.2 Previous Work

While there is a plethora of published reports regarding eye movements by drivers and their perception of various road features, including road markings, they were done mostly in laboratory [5]. A notable experimental field work confirming seven types of gazing strategies was done only in rural environment [6]. Results from field analyses combining horizontal road markings and approaches to intersections, done in an urban environment, with all of the distractions to which drivers are constantly exposed, apparently were not published before the work described in this experiment [7]. The advantage of the utilised experimental procedure is the possibility of recording the real-world scenarios, but highly disadvantageous is the lack of possibility to record responses to rare events, which can be conveniently simulated in laboratory.

In the first article related to the perception of horizontal signalisation by drivers in urban area, analysed was the far distance, 70–180 m from the intersection [8]. It was confirmed that the perception was mostly at the level of gazes and that a fixation, if it occurred, was relatively long. In the subsequent report, visual distances of 35–70 m and 70–180 m were compared in terms of gazes and it was discovered that the width of intersection had profound influence on the distance, at which gazing at road markings occurred [9]. In this article presented are further analyses from the same test area, including the influence of drivers' experience and sex. We believe that deeper understanding of the drivers' eye movements might permit for better design of road and road markings and in consequence increase the comfort of driving and safety.

1.3 Eye Tracking

Eye tracking is an established technique recording eye movements in response to a stimulus. It is a frequently used tool for analysis of drivers' behaviour and perception, including analysis of horizontal signalisation [7]. Typically, four main parameters are recorded: gaze (a single eye regard at a specific position), fixation (a group of gazes within a defined space), fixation duration, and saccade (a change of eye focus between fixations) [10]. Measurements of physiological responses, including pupil diameter is also done sometimes to provide supplementary information about stress.

2 Experimental

2.1 Test Area and Test Participants

The study was done within city limits of Kraków, Poland, in an area comprising high-rise residential blocks of low-end flats separated by dual carriageway streets. The 7.9 km route was taken on roads that at intersection approaches are broadened to even five lanes. Amongst the route features one must list roundabouts, turnarounds, pedestrian crossings (both controlled and not controlled by traffic lights), public transport infrastructure for trams and buses, and occasional bicycle paths. A map with the route and the analysed intersections marked is shown in Fig. 1. Like in the previous work, intersections 3, 7, and 10 were not analysed because of their specific arrangement [8, 9].

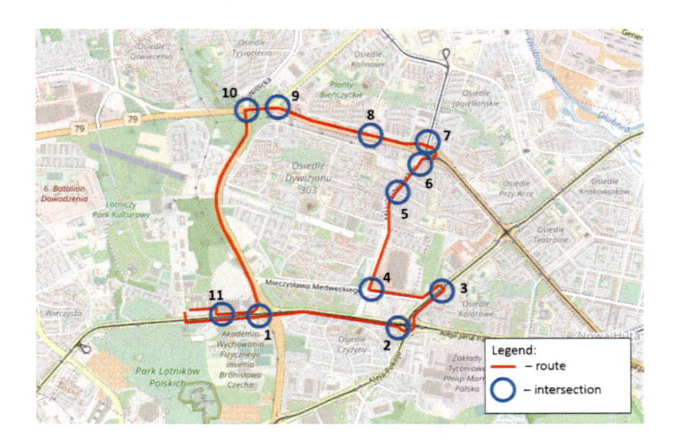

Fig. 1. Map of the test route with the intersections marked.

The tested drivers were recruited amongst university students and were selected after filling a standard questionnaire regarding their basic demographic information, vision, and driving experience. They were all volunteers, not compensated for their work. A total of 32 participants, all holding valid category B driving licence and having corrected or uncorrected 6/6 vision were selected. Amongst them, only 4 declared no knowledge of the area, which we must acknowledge as having a potential to influence the experiment's results. Ethical guidelines set by Politechnika Krakowska were followed at all stages of the experiment. Basic information about the participants is shown in Table 1.

Table 1. Test participants.

Participants	Number	Age [years][a]	Average distance driven per month		
			Below 200 km	200–1000 km	Over 1000 km
All	32	24.1 (1.7)	23 (72%)	6 (19%)	3 (9%)
Male	21 (65%)	24.4 (2.0)	14 (67%)	4 (19%)	3 (14%)
Female	11 (35%)	24.2 (1.0)	9 (82%)	2 (18%)	–

[a]Standard deviations are given in parentheses.

2.2 Equipment and Data Analysis

The eye tracking device FOVIO (Seeing Machines; Canberra, Australia) was mounted on the steering column of a 2014 model of a compact car powered by 1.25 dm^3 petrol engine and equipped with a 5-speed manual transmission. The test participants were given a brief time to familiarize themselves with the vehicle, during which time the performance of eyetracker device was verified and the equipment was calibrated. Drivers were not informed earlier about the planned travel path; they were instructed to obey speed limits and other rules of the road. Verbal instructions regarding the driving path were given by the research assistant present in the passenger seat. Tests were run in late summer, with weather conditions varying from sunny to cloudy. Driving was done during off-peak hours on work days.

Data collection and analysis was done using EyeWorks™ software (EyeTracking Inc.; Solana Beach, California, U.S.A.). There were quite notable data voids, while drivers moved their head, but this did not affect the overall quality of data. Even though the recordings were continuous, only the regions of intersections' approaches were analysed for the purpose of this report.

3 Results and Discussion

3.1 Visualisation with Heat Maps

For visual presentation, heat maps comprising overlay of all of the results were selected as the simplest and most readable method for data overview [11]. Heat maps of the analysed intersections and visual regions are shown in Table 2.

3.2 Observations, Fixations, and Gazes

In Table 3 are provided the collected data for the numbers of observations, fixations, and gazes for both far and near visual distances. It can be noticed that the observation rate of horizontal road markings was very low: only $\pm 10\%$ of drivers fixated their sight and only 17% gazed. The small number of recorded observations can serve as a confirmation that the horizontal signalisation is used only for confirmation of position, and not as the main visual cue. It is likely that during driving outside of the city, particularly on unlit roads, road markings would take much more prominent role [3]. The difference between far and near visual distances were small and statistically not significant. Fixations duration, averaging 0.17 s (standard deviation 0.07 s), was also similar regardless of the visual distance. The average numbers of gazes, 32 at the far distance and 28 at the near distance, appeared to be more different, but the difference was statistically the same as in case of fixation durations.

The intersections were then separated based on their size into broad approach with four or five traffic lanes (1, 2, and 6), narrow approach with one or two traffic lanes (4 and 5), and typical approach with three traffic lanes (8, 9, and 11). The results, given in Table 4, clearly demonstrate that horizontal road markings at the approaches to broad intersections were observed, both at the level of fixation and glances, in the far visual field and the opposite held true for narrower approaches to intersections. This may serve as another confirmation of the guiding role of road markings.

3.3 Influence of Driver Sex and Experience

In the work related to perception of horizontal road markings with high retroreflectivity, differences between sexes were found, with female drivers reporting observing road safety features more seldom than male [12]. However, analysis of the drivers based on their experience, demonstrated that the differences were based not on sex of

Table 2. Heat maps (accumulated overlay for all drivers).

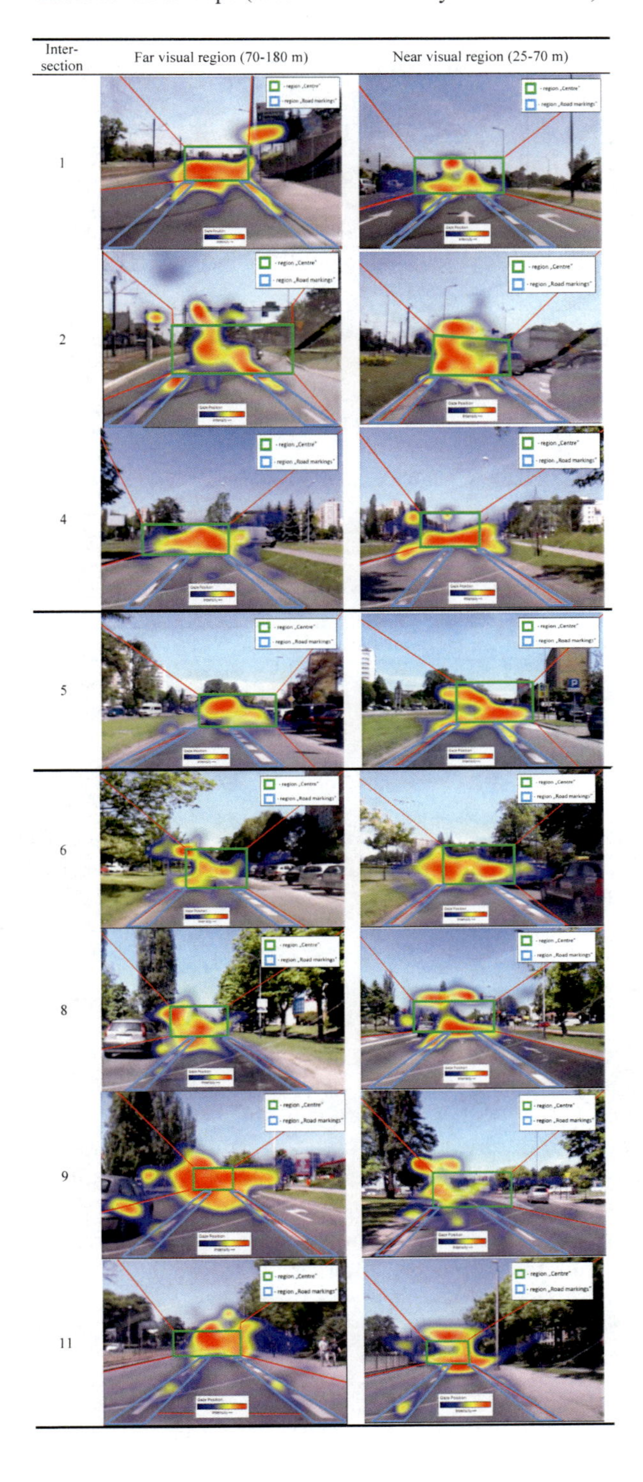

Inter-section	Far visual region (70-180 m)	Near visual region (25-70 m)
1		
2		
4		
5		
6		
8		
9		
11		

Table 3. Observations, fixations, and gazes at horizontal road markings.

Intersection	Fixations				Fixation duration [s]		Gazes			
	Observers		Average number				Observers		Average number	
	Far	Near	Far	Near	Far	Near	Far	Near	Far	Near
1	6	5	3.8	1.8	0.25	0.22	10	9	49	23
2	3	1	6.3	1.0	0.19	0.17	8	3	56	7
4	2	1	1.0	2.0	0.12	0.13	3	1	11	29
5	2	3	1.5	3.3	0.10	0.20	2	5	31	45
6	2	1	1.0	1.0	0.37	0.10	3	4	15	9
8	4	4	3.3	2.5	0.20	0.19	10	5	31	37
9	2	2	1.5	2.0	0.18	0.13	4	3	17	33
11	6	5	1.5	1.4	0.13	0.18	8	7	49	40

Table 4. Observations, fixations, and gazes at horizontal road markings, divided by intersection size.

Intersections	Fixations				Fixation duration [s]		Gazes			
	Observers		Average number				Observers		Average number	
	Far	Near	Far	Near	Far	Near	Far	Near	Far	Near
1, 2, 6	11	7	4.0	1.6	0.23	0.20	17	11	58	24
4, 5	2	3	2.5	4.0	0.11	0.18	3	5	32	51
8, 9, 11	9	7	2.8	3.0	0.17	0.18	14	9	54	62

the driver but driving experience; data was skewed due to lesser number of kilometres driven by the average questioned female driver. The same type of analyses was applied to results presented herein.

Firstly, the results were separated based on sexes, as shown in Table 5. One ought to note that the fixations and gazes were recorded only for some of the drivers, so for the presented data one should consider that there were 21 males and 11 females. On average, the differences are insignificant when it comes to the near and far distances at the same intersections; however, in cases of female drivers the number of gazes was larger (34 as compared to average 21 for male). Average fixation time was also slightly larger for females (0.23 s) as compared to males (0.17 s).

The cross-analysis of drivers' sex and size of intersection, shown in Table 6, demonstrates the same pattern that was analysed for all drivers, but with male drivers being less likely to fixate or gaze at the horizontal road markings than females. Next, the assessment was extended to drivers' experience, based on average number of kilometres driven monthly. Amongst observers, there were 23 drivers who drive less than 200 km monthly (classified in the text as inexperienced) and 9 drivers who drove

Table 5. Observations, fixations, and gazes at horizontal road markings, divided by sex of the drivers.

Intersection	Fixations				Fixation duration [s]		Gazes			
	Observers		Average number				Observers		Average number	
	F	M	F	M	F	M	F	M	F	M
1 far	4	2	3.0	5.5	0.29	0.20	5	5	44	53
1 near	3	2	2.3	1.0	0.19	0.30	4	5	32	15
2 far	1	2	5.0	7.0	0.16	0.20	3	5	41	65
2 near	1	0	1.0	–	0.17	–	2	1	5	10
4 far	2	0	1.0	–	0.12	–	2	1	14	6
4 near	1	0	2.0	–	0.13	–	1	0	29	–
5 far	2	0	1.5	–	0.10	–	2	0	31	–
5 near	3	0	3.3	–	0.20	–	5	0	45	–
6 far	2	0	1.0	–	0.37	–	2	1	21	4
6 near	0	1	–	1.0	–	0.10	2	2	14	3
8 far	3	1	4.0	1.0	0.21	0.10	6	4	43	12
8 near	4	0	2.5	–	0.19	–	5	–	37	–
9 far	1	1	2.0	1.0	0.12	0.29	1	3	32	11
9 near	1	1	2.0	2.0	0.14	0.12	1	2	19	40
11 far	5	1	1.6	1.0	0.14	0.10	5	3	65	22
11 near	3	2	1.7	1.0	0.19	0.16	4	3	52	24

Table 6. Observations, fixations, and gazes at horizontal road markings, divided by sex of the drivers and by intersection size.

Intersections	Fixations				Fixation duration [s]		Gazes			
	Observers		Average number				Observers		Average number	
	F	M	F	M	F	M	F	M	F	M
1, 2, 6 far	7	4	2.7	6.3	0.26	0.20	7	10	55	59
1, 2, 6 near	4	3	2.0	1.0	0.19	0.23	5	6	33	16
4, 5 far	2	0	2.5	–	0.11	–	2	1	45	6
4, 5 near	3	0	4.0	–	0.18	–	5	0	51	–
8, 9, 11 far	6	3	3.7	1.0	0.17	0.17	7	7	88	21
8, 9, 11 near	4	3	4.3	1.3	0.19	0.14	5	4	82	38

more (classified herein as experienced). The results provided in Table 7 show dramatic difference depending on drivers' experience. Inexperienced drivers were much more likely to focus on the markings (on average 2.8 as compared to 0.3 for experienced) and to gaze at them (average 4.5 as compared to 0.8 for experienced). However, the number of gazes, if they occurred, was similar regardless of the experience. Results based on the intersection size, given in Table 8, do not depart from previously established pattern of earlier gazes at horizontal signalisation ahead of larger intersections.

Table 7. Observations, fixations, and gazes at horizontal road markings, divided by average distance driven monthly (below or above 200 km).

Intersection	Fixations				Fixation duration [s]		Gazes			
	Observers		Average number				Observers		Average number	
	<200	>200	<200	>200	<200	>200	<200	>200	<200	>200
1 far	5	1	3.6	5	0.19	0.43	9	1	40	126
1 near	4	1	2.0	1	0.21	0.29	7	2	27	10
2 far	2	1	7.0	5	0.20	0.16	7	1	52	85
2 near	1	0	1.0	–	0.17	–	2	1	5	10
4 far	2	0	1.0	–	0.12	–	2	1	14	6
4 near	1	0	2.0	–	0.13	–	1	0	29	–
5 far	2	0	1.5	–	0.10	–	2	0	31	–
5 near	3	0	3.3	–	0.20	–	4	1	49	30
6 far	2	0	1.0	–	0.37	–	2	1	21	4
6 near	1	0	1.0	–	0.10	–	4	0	9	–
8 far	4	0	3.3	–	0.20	–	8	2	33	23
8 near	4	0	2.5	–	0.19	–	5	0	37	–
9 far	2	0	1.5	–	0.18	–	3	1	17	15
9 near	2	0	2.0	–	0.13	–	3	0	33	–
11 far	5	1	1.4	2.0	0.12	0.16	6	2	61	12
11 near	5	0	1.4	–	0.18	–	7	0	40	–

Table 8. Observations, fixations, and gazes at horizontal road markings, divided by average distance driven monthly and by intersection size.

Intersections	Fixations				Fixation duration [s]		Gazes			
	Observers		Average number				Observers		Average number	
	<200	>200	<200	>200	<200	>200	<200	>200	<200	>200
1, 2, 6 far	9	2	3.8	5.0	0.21	0.30	14	3	55	72
1, 2, 6 near	6	1	1.7	1.0	0.19	0.29	9	2	26	15
4, 5 far	2	0	2.5	–	0.11	–	2	1	45	6
4, 5 near	3	0	4.0	–	0.18	–	4	1	56	30
8, 9, 11 far	8	1	2.9	2.0	0.17	0.16	11	3	62	28
8, 9, 11 near	7	0	3.0	–	0.18	–	9	0	62	–

4 Discussion and Conclusions

Expectedly, the pavement markings were mostly perceived at the level of gazes and such observations were recorded only in cases of only 10% of the test participants. The recorded fixation periods were relatively long (average 0.18 s, as compared with average focus time at other areas of 0.24–0.30 s). Subsequent analysis revealed that it was most likely caused by inexperience of the drivers. The distance at which gazes at road markings occurred depended on the size of the oncoming intersection. Large intersections caused earlier gazes, which can be understood as a visual search to appropriately position the vehicle. The pattern held true regardless of the experience of the drivers. While gazes were more frequently recorded for less experienced drivers, their number was similar regardless of drivers' experience. The use of gazes instead of focus suggest that the markings were perceived as vehicle positioning confirmation signals. Significantly more frequent observations of horizontal road markings by inexperienced drivers serves as a validation that that they seek additional cues. It was not possible to verify whether knowledge of the area could have influenced the behaviour. Amongst methodological issues that could have influenced the outcome, the most important is the use of static eyetracker, which led to significant data voids when the drivers moved their head or changed position. Data quality report was obtained from the software and deemed acceptable for data processing [13].

Since this was a preliminary study aimed at verification whether there would be valid differences worth further exploration, significant additional research has to be proposed. Firstly, a significant limitation of this experiment was the exclusion of experienced and elderly drivers who were reported to rely more frequently on horizontal signalisation than on dynamic cues [14]. Testing at two-lane roads and assessments at different time of the day, particularly at night, should also be performed. For night time driving, retroreflectivity of road markings is of critical importance and in this field there are novel developments that have a strong potential of influencing the eye movements of all drivers [15]. Finally, it must be added that the presence and clarity of horizontal road markings is critical not only for human drivers, but also for machine vision and the emerging technology of autonomous vehicles that rely on it [16]. Field analysis done simultaneously on machine vision equipment and human drivers could be of profound importance.

References

1. World Health Organisation: Global status report on road safety 2015. http://www.who.int/violence_injury_prevention/road_safety_status/2015/GSRRS2015_Summary_EN_final2.pdf?ua=1. Last accessed 18 Nov 2017
2. Makarova, I., Shubenkova, K., Mukhametdinov, E., Pashkevich, A.: Safety related problems of transport system and their solutions. In: Proceedings of the 11th International Science and Technical Conference Automotive Safety, Častá - Papiernička, Slovakia, 18–20 April 2018. https://doi.org/10.1109/AUTOSAFE.2018.8373333
3. Steyvers, F.J., De Waard, D.: Road-edge delineation in rural areas: effects on driving behaviour. Ergonomics **43**(2), 223–238 (2000)
4. Miller, T.R.: Benefit–cost analysis of lane marking. Transp. Res. Rec. J. Transp. Res. Board **1334**, 38–45 (1992)
5. Lemonnier, S., Brémond, R., Baccino, T.: Gaze behavior when approaching an intersection: dwell time distribution and comparison with a quantitative prediction. Transp. Res. Part F: Traffic Psychol. Behav. **35**, 60–74 (2015)
6. Lappi, O., Rinkkala, P., Pekkanen, J.: Systematic observation of an expert driver's gaze strategy—an on-road case study. Front. Psychol. **8**, 620 (2017)

7. Taylor, T., Pradhan, A.K., Divekar, G., Romoser, M., Muttart, J., Gomez, R., Pollatsek, A., Fisher, D.L.: The view from the road: the contribution of on-road glance-monitoring technologies to understanding driver behavior. Accid. Anal. Prev. **58**, 175–186 (2013)
8. Pashkevich, A., Burghardt, T.E., Kubek, D.: Analysis of drivers' eye movements ahead of intersections: observations of horizontal road markings. In: Proceedings of International Conference on Science and Traffic Development, Opatija, Croatia, pp. 201–208, 10–11 May 2018
9. Pashkevich, A., Burghardt, T.E., Kubek, D.: Drivers' gazes at horizontal road markings ahead of intersections. In: Proceedings of International Conference on Traffic and Transport Engineering, Belgrade, Serbia, pp. 849–853, 27–28 September 2018
10. Raschke, M., Blascheck, T., Burch, M.: Visual analysis of eye tracking data. In: Huang, W. (ed.) Handbook of Human Centric Visualization, pp. 391–409. Springer, New York (2014)
11. Špakov, O., Miniotas, D.: Visualization of eye gaze data using heat maps. Elektronika ir elektrotechnika **2**(74), 55–58 (2007)
12. Burghardt, T.E., Pashkevich, A., Piegza, M.: Drivers' perception of horizontal road marking with high retroreflectivity. Transport Miejski i Regionalny **8**, 5–10 (2017). (in Polish)
13. Holmqvist, K., Nyström, M., Mulvey, F.: Eye tracker data quality: what it is and how to measure it. In: Proceedings of the Symposium on Eye Tracking Research and Applications, Santa Barbara, California, pp. 45–52, 28–30 March 2012
14. Dukic, T., Broberg, T.: Older drivers' visual search behaviour at intersections. Transp. Res. Part F: Traffic Psychol. Behav. **15**(4), 462–470 (2012)
15. Burghardt, T.E., Pashkevich, A., Fiolić, M., Żakowska, L.: Horizontal road markings with high retroreflectivity: durability, environmental, and financial considerations. Adv. Transp. Stud. Int. J. **47**, 49–60 (2019)
16. Carlson, P.J., Poorsartep, M.: Enhancing the roadway physical infrastructure for advanced vehicle technologies: a case study in pavement markings for machine vision and a road map toward a better understanding. In: Proceedings of Transportation Research Board 96th Annual Meeting, Washington, DC, 8–12 January 2017

Safety Analysis of Road Networks in Germany – Approaches of Section Development and Comparison to Other Countries

Johannes Vogel(✉), Julius Uhlmann, and Uwe Plank-Wiedenbeck

Bauhaus-Universität Weimar, Chair of Transport System Planning, Weimar, Germany
{johannes.vogel, julius.uhlmann, uwe.plank-wiedenbeck}@uni-weimar.de

Abstract. To meet the principles of the European Directive on Road Infrastructure Safety Management, Germany has established the Safety Analysis of Road Networks. With this reactive method, the safety of the road network in operation can be evaluated. Aim of the analysis is to get an overview of the traffic safety of the road network and to identify road sections which feature a high potential to improve traffic safety. An improvement of the infrastructure in these sections is supposed to be especially important and also cost-effective. In which way the redesign of the infrastructure can improve traffic safety of a section is not part of the analysis, but has to be investigated more in detail [1].

For that purpose, the road network to be analysed has to be divided into comparable sections. The German Recommendations for the Safety Analysis of Road Networks (ESN) offer two options for section development: either by accident structure or network structure [2]. A third option is the so-called integral method, which is not included in the guidelines, though [3].

After determining the sections, a so-called safety potential is calculated for each section. It is defined as the accident costs per kilometre, which could be prevented by a road design in accordance with the relevant guidelines. After the calculation, the sections can be ranked by their safety potential. The sections with the highest safety potentials should be investigated more in detail, e.g. by an extensive analysis of the accident data [2].

Aim of the paper is to discuss advantages and disadvantages of the different options of section development and to compare the German approach with other approaches.

Keywords: Safety analysis of road networks · Safety potential · Section development · Integral method

1 Method of Safety Analysis of Road Networks

1.1 Purpose and Character of Safety Analysis

The safety analysis of road networks in Germany is based on the "Recommendations for the Safety Analysis of Road Networks" (short: ESN) [2]. They were developed and published by the Road and Transportation Research Association (FGSV) in order to

A. Varhelyi et al. (Eds.): VISZERO 2018, LNITI, pp. 11–20, 2020.
https://doi.org/10.1007/978-3-030-22375-5_2

meet the demand for a Network Safety Management, which was required by the EU Directive on Road Infrastructure Safety Management (2008/96/EC).

The purpose of the safety analysis is to give an overview of the traffic safety in road networks in operation. It can be used to determine, where in the road network safety deficits in the sense of many and severe traffic accidents exist. Therefore it can give important indications about possible deficits in road routing, design and conditions. Hence it is no substitute for a thorough analysis of accidents, but can be combined e.g. with a black spot analysis or a road safety inspection [2].

1.2 Safety Potential

The central element of the safety analysis is the so-called safety potential (SAPO). It is characterized as the accident costs per kilometer [1,000 €/(km*a)], which could be avoidable in case of a road design in accordance with the relevant guidelines (accident cost density ACD). It is calculated from the difference between the actual incurred ACD and the basic accident cost density bACD (see Formula 1 and Fig. 1) [2].

$$SAPO = ACD - bACD \tag{1}$$

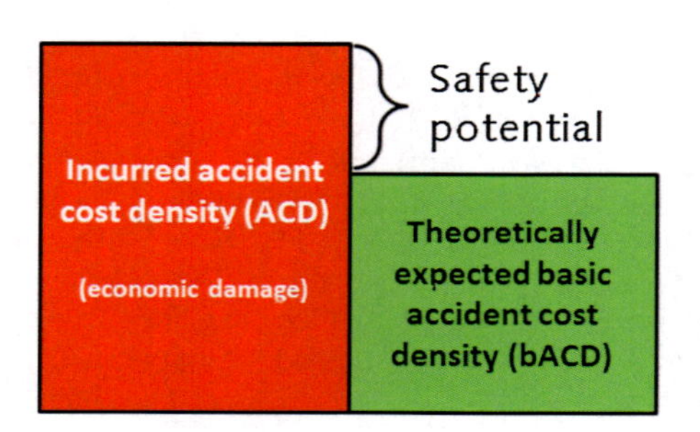

Fig. 1. Calculation of the safety potential [3].

The bACD are those accident costs, which are not influenced by the infrastructure anymore, but only by the components vehicle and human [1]. Hence, from the infrastructure point of view, they are unavoidable and therefore theoretically expected [3].

1.3 Procedure of Safety Analysis

To calculate the safety potential, the average daily traffic and the accident data (number, severity and location of accidents) for the entire road network are required [3]. Even though the data availability is increasing, unfortunately there is still often a lack of traffic volume data, which has to be compensated by model values [4]. This is one issue of the safety analysis, which still has to be improved.

After data acquisition, the road network has to be divided into reasonable and comparable sections. The development of these sections can be done either by accident structure, network structure or the so-called integral method [3]. The three approaches will be explained more in detail in the following chapter.

Afterwards, for each of these sections the safety potential can be calculated. To give an overview of the traffic safety of the road network, the sections can be prioritized and mapped with different colors according to their SAPO [2–4].

2 Approaches of Section Development

2.1 Section Development by Accident Structure

Sectioning the road network by accident structure means to develop the sections by visual accident density. On the basis of a map showing the accidents with serious personal injuries within three years, contiguous road parts with more or less the same accident density are defined as one section. It's also possible to define additional section borders in case of a major change in traffic volume. According to the ESN, a section with less than three accidents with serious personal injuries should be merged with a neighboring section. Accidents at junctions shall be assigned to the section with the highest accident density of accidents with serious personal injury [2]. Moreover, it is important to note that the section development is based on accidents with serious personal injuries, but the SAPOs are calculated with all accidents, including with light personal injuries and property damage only [5].

After developing the sections, for each section the ACD, bACD and SAPO are calculated. The sections then can be mapped in different colors, according to their SAPO. A fictitious example with the results of safety potential calculation is shown in Fig. 2. The figure should be read section by section from bottom to top: In the first

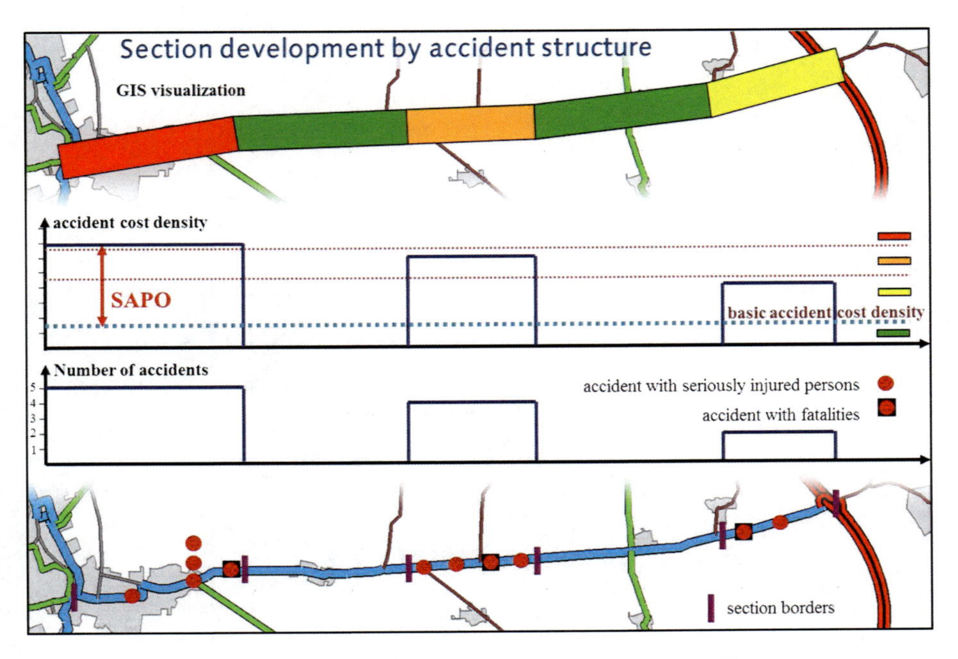

Fig. 2. Section development by accident structure [5].

section there are four accidents with a very high ACD, which results in a high SAPO. Therefore the section is mapped with red color etc. [5].

The advantage of this approach is the scientific accuracy due to the basis of incurred accidents. Sections with many and/or severe accidents are reliably assigned a high safety potential [5].

Disadvantages are the high working effort for the analysis of a whole road network and the high subjectivity of this approach. Depending on the person in charge, the section borders can be set quite differently. Moreover, calculations from different periods of time are not comparable with each other, because the section borders might vary due to a different amount and location of the accidents [5].

The results of this approach show a clear and easily interpretable picture, even though the informative value is limited by the influence of short sections: If these sections also feature few but severe accidents (serious personal injuries or fatalities), the ACD is rising. This results in a higher SAPO, although this is only due to the short section length [5, 6].

2.2 Section Development by Network Structure

The approach of section development by network structure uses junctions, relevant changes of the infrastructure (e.g. cross sections) or the average daily traffic to set section borders. The sections should be developed on the basis of a network map including junctions and town borders as well as on a traffic volume map [5].

Figure 3 shows the same fictitious example like before. This time, though, the sections were developed by network structure.

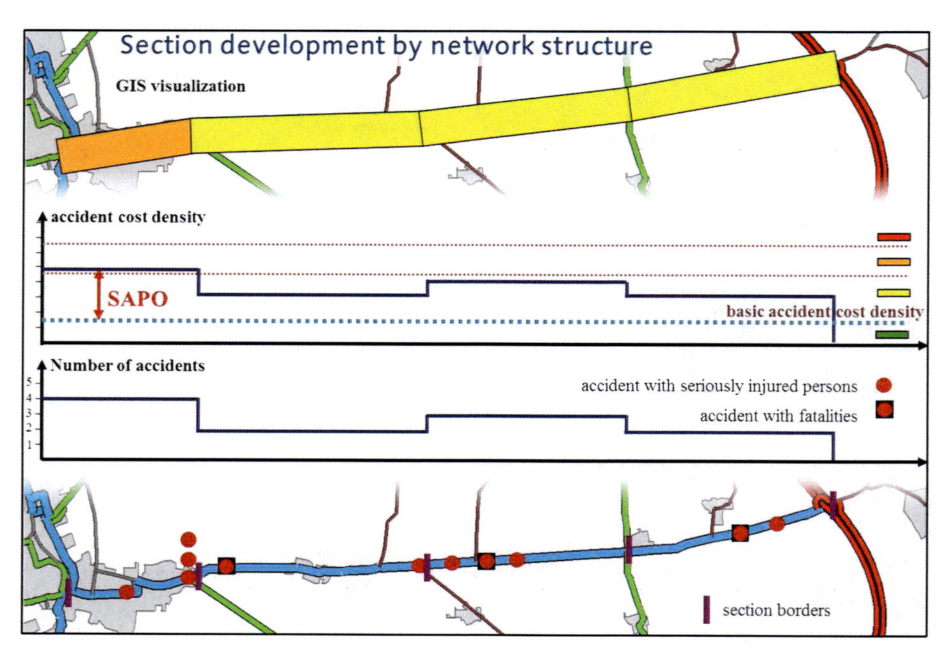

Fig. 3. Section development by network structure [5].

The advantage of this approach is that – in contrast to the first approach – calculations of different time periods can be compared. Junctions usually do not change that often and therefore the section borders are mostly the same [5].

Disadvantages are as well the high working effort for the analysis of a whole road network and the high subjectivity of this approach. In addition, detailed information about the road network (e.g. exact position of the junctions and changes of the cross sections) are necessary, which may be difficult to get [5].

Similar to the section development by accident structure, the results of this approach show a clear and easily interpretable picture, but their informative value is limited by the influence of short sections (see explanation above) [5, 6].

2.3 Section Development by Integral Method

The section development by the integral method works with a gliding segment (integral) with defined width and step size, which is moved across the entire network. Hence, this approach works independently from accidents, junctions, infrastructure and traffic volume [3, 5].

The difference to the other two approaches is that the sections are overlapping each other. Since the integral width (e.g. 1,000 m) is usually a multiple of the step size (e.g. 100 m), each section overlaps with several others. Therefore in the GIS visualization it is not possible to display the whole section. Instead, only the step size starting from the middle of the section is depicted (see Fig. 4). It is important to know that the displayed color applies not only for the shown width (step size), but for the whole integral width [3].

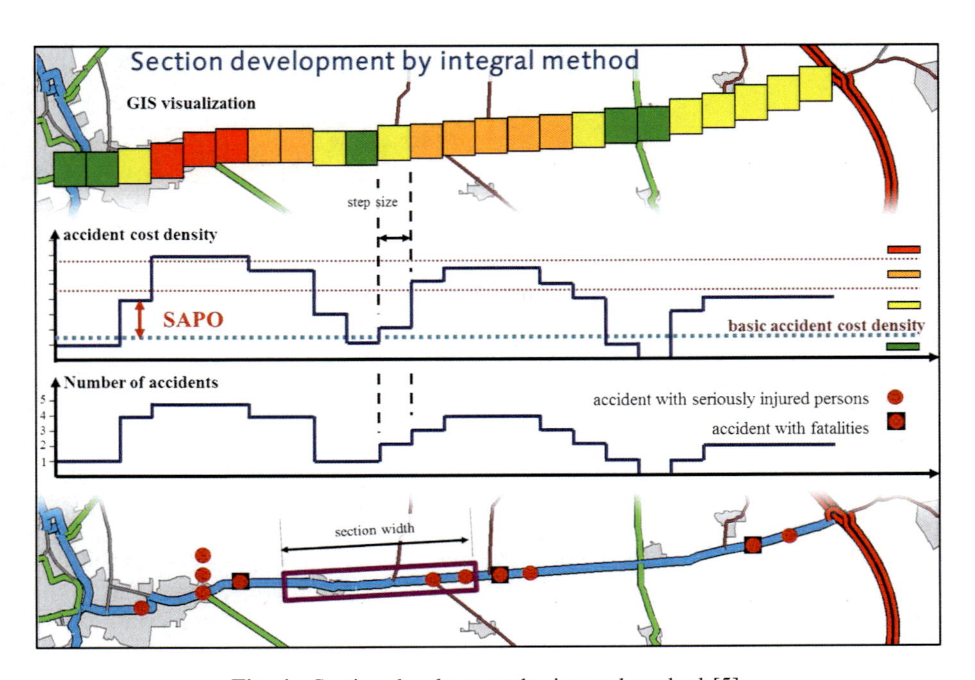

Fig. 4. Section development by integral method [5].

The advantages of this approach are the objectivity – there are no subjectively defined section borders – and the easy handling because it is a computer-based method. It is also possible to compare calculations of different time periods because the section borders do not change [3, 5].

One big disadvantage is that the GIS visualization can be misinterpreted as explained above: the calculated safety potential applies for the whole section width, even though it is only displayed for the width of the step size. Also, due to the gliding segment (integral) which is moved across the network, different safety potentials ("running SAPOs") can be found in between apart from that reasonable section borders (see e.g. the road part from section seven (orange) to section eleven (yellow) in Fig. 4). The approach of the integral method is also less descriptive and therefore much more difficult to understand. Moreover calculation parameters (integral width, step size) have to be defined in advance according to the network type (motorways, federal highways, state roads etc.). Last but not least, the approach is not yet included in the guidelines [5].

2.4 Comparison and Discussion

As Fig. 5 depicts, the three approaches of section development show inevitably similar, but different results. This confirms that the choice of the section development approach has a major influence on the calculated safety potential [5]. It also reflects the results of various research studies [3, 6].

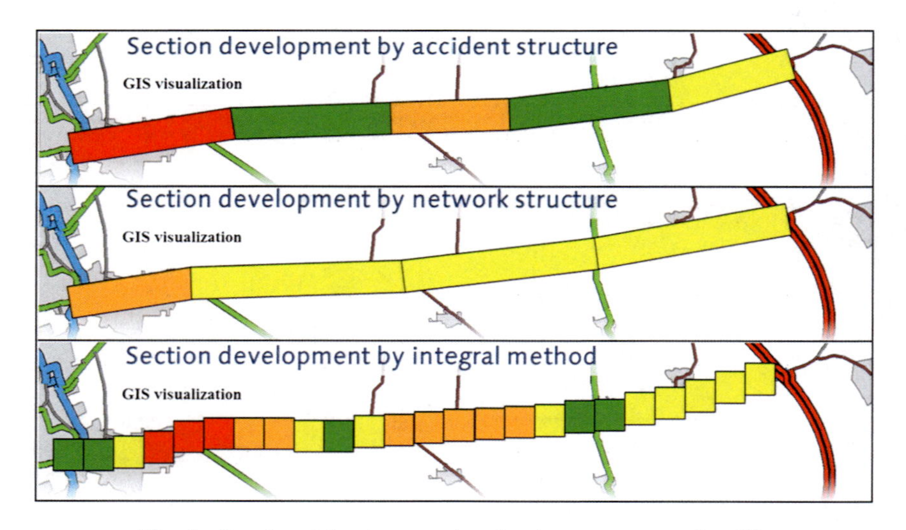

Fig. 5. Results of the three section development approaches [5].

Table 1 lists the advantages and disadvantages of the three approaches again. It is visible that all methods have a similar number of benefits and drawbacks. None of the approaches can be clearly defined as best. It is obvious for example that the approaches based on network structure and integral method are more appropriate for a comparison

Table 1. Advantages and disadvantages of the three section development approaches [5].

	Section development by...		
	Accident structure	Network structure	Integral method
Advantages	– Scientific accuracy – Easily interpretable	– Comparability between different time periods – Easily interpretable	– Less working effort – Objectivity – Comparability between different time periods
Disadvantages	– High working effort – High subjectivity – Not comparable to other time periods – Limited informative value due to influence of short sections	– High working effort – High subjectivity – Necessity of detailed information – Limited informative value due to influence of short sections	– Possible misinterpretation of the GIS visualization – "Running SAPOs" in between otherwise reasonable section borders – Definition of calculation parameters beforehand – Not included in ESN

between safety analyses of different time periods. If, on the other hand, an easy understanding and interpretation of the results are more important, then the approaches by accident or network structure are more convenient. Therefore the decision for the method to use has to be taken on an individual basis.

For the State Road Network in Thuringia (Germany), the safety analysis is carried out every three years by the Chair of Transport System Planning of Bauhaus University Weimar. So far, the network was analyzed in 2013 and 2016, both times with the integral method of section development. The reason for this decision was the fact that this approach is more objective than the others and that the specific software program, which is required for the calculation, was available at the State Office for Construction and Transport. Therefore the integral method was also much easier to handle. The disadvantage of a possible misinterpretation was attempted to counteract with intensive explanation and information about the approach and its results. Furthermore, also the continuous performance of the analysis every three years contributes to the compensation of possible disadvantages and provides important and comparable results and findings [4, 5]. For instance, the results of the 2016 analysis are an important part of the State Road Plan 2030 for Thuringia [7].

3 Approaches of Section Development in Other Countries

3.1 Background

Against the background that the integral method seems to work well, but is not yet included in the guidelines, it was investigated, which methods are used abroad. It was found that more or less similar procedures to the integral method exist among others in Norway and especially in the US. Since these two approaches are documented quite well, they will be briefly presented below.

3.2 Section Development in Norway

The Norwegian approach to section development for safety analysis of road networks is quite clear and strict. All national roads are divided into 1-km sections. For each section, the accident data and several explanatory variables (average daily traffic, speed limit, road type, number of lanes, number of junctions per kilometer, whether the road has the status of a national main road or not) are needed. Sections with missing or discontinuous data on any explanatory variables are not taken into account [8].

For each section, the so-called expected Injury Severity Density (ISD) can be calculated. It is defined by the sum of the weighted numbers for killed, critically, seriously and slightly injured road users per kilometer and year. To estimate the expected ISD, the empirical Bayes method is applied. It combines the estimated and the recorded number of injured road users by multivariate models for each road section [8].

The safety of the road sections can then be determined on the basis of the expected ISD. For this, the national roads are divided into three classes. "Green roads" are the safest roads (lowest 50% of expected ISD values), where also no fatalities or seriously injured persons were recorded. "Red roads" are the most dangerous roads (highest 10% of expected ISD values), where fatalities or seriously injured persons were recorded. "Yellow roads" are the remaining 40% of the roads, which neither belong to the "red" nor to the "green" roads [8].

After the most hazardous road sections have been identified with that method, an additional accident analysis is carried out to determine appropriate safety measures which can help reducing the ISD [8].

3.3 Section Development in the United States

In the US, available data on road characteristics (geometric design and traffic control features, traffic volumes) and road safety (accident history and characteristics etc.) are used to identify and prioritize the most hazardous road sections [8].

The road network therefore is separated into homogenous road sections according to the data on the road characteristics. Each section is then divided into 0.1-mile subsections. Subsequently, a window of X consecutive subsections is placed with its left edge at the left margin of the homogenous road section and is moved subsection by subsection to the right. This is done until its right edge reaches the right margin of the homogenous road section and is then repeated for all feasible sizes of X [8].

For each subsection, the expected accident frequency (EAF, accidents per mile and year) is calculated with the available accident data. From these results, the average EAF for each status of the window (so-called segment) is computed (see Fig. 6). The highest average so calculated represents the highest peak for window size X. In Fig. 6 for instance segment B is the highest peak of window size three. In the case of two overlapping segments of different size (B and C), only the one with the higher average is considered further (B). Finally, all peaks and subsections to be considered are ranked for a more detailed investigation regarding their prospective cost-effectiveness of potential safety measures [8].

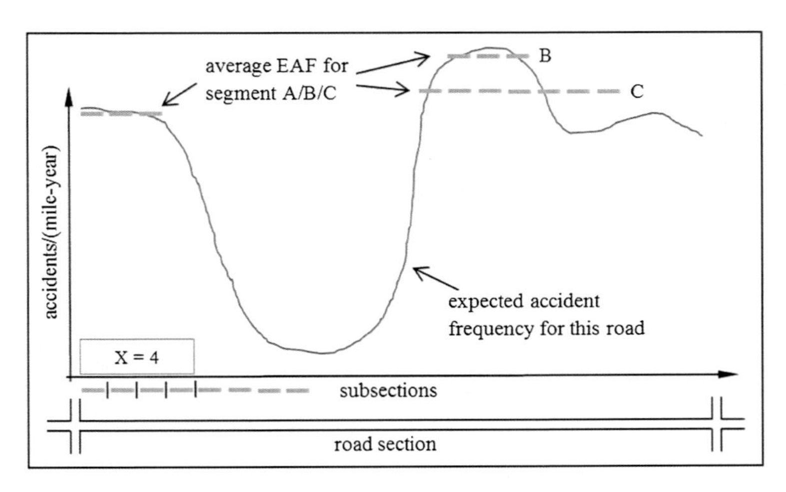

Fig. 6. Safety analysis approach used in the US [8].

4 Conclusion and Outlook

The previous chapters have shown that the way in which the road sections are formed has a high influence on the results of the safety analysis. This also was confirmed by various research studies in the past [3, 6].

All three approaches used for the safety analysis in Germany have several advantages and disadvantages, which need to be considered beforehand. The approach by accident structure stands for scientific accuracy and good comprehensibility, but also for high workload and subjectivity as well as lack of temporal comparability. The approach by network structure is characterized by the possibility of temporal comparability and good comprehensibility, but also by high workload and subjectivity as well as by the influence of short road sections. Typical for the integral method is a lower workload and good temporal comparability, but also a difficult interpretation and varying SAPOs in between otherwise reasonable section borders [5].

In general, the safety analysis is a good and important application to get an overview of the traffic safety of road networks. However, there is still room for improvement, in particular regarding the available data for traffic volumes and the development of road sections.

Moreover, among others in Norway and especially in the US procedures for section development are used that are much more similar to the integral method than to the other German approaches [8]. Therefore, it is proposed to evaluate the experiences with these approaches in Norway and the US to derive insights for the German procedure. It also should be investigated again more detailed, if the integral method should not at least be included in the German guidelines. This could increase the acceptance and importance of this approach significantly so that it would be considered at least an equivalent procedure for section development to the two other methods.

References

1. Chambon, P., Ganneau, F., Lemke, K.: Sicherheitsmanagement von Straßennetzen – Eine französisch-deutsche Zusammenarbeit. In: TRA – Transport Research Arena Europe 2006: Goeteborg, Sweden, June 12th–15th 2006: Greener, Safer and Smarter Road Transport for Europe. Proceedings. Swedish Road Administration, Goeteborg (2006)
2. Forschungsgesellschaft für Straßen- und Verkehrswesen: Empfehlungen für die Sicherheitsanalyse von Straßennetzen ESN. Köln (2003)
3. Brannolte, U., Grießbach, A., Viehmann, I.: Sicherheitsanalyse von Straßennetzen nach ESN – fachliche Untersuchung der Integralen Methode. Erläuterungsbericht. Professur Verkehrssystemplanung an der Bauhaus-Universität Weimar, Weimar (2012)
4. Grießbach, A., Plank-Wiedenbeck, U., Vogel, J.: Sicherheitsanalyse des Landesstraßennetzes Thüringen – Ergebnisbericht. Professur Verkehrssystemplanung an der Bauhaus-Universität Weimar, Weimar (2017)
5. Grießbach, A., Plank-Wiedenbeck, U., Vogel, J.: Verkehrssicherheit bei Planung, Entwurf und Betrieb – Sicherheitsanalyse von Straßennetzen. Interne Vorlesungsunterlagen der Professur Verkehrssystemplanung an der Bauhaus-Universität Weimar, Weimar (2018)
6. Ebersbach, D., Schüller, H.: Praktische Anwendung der Empfehlungen für die Sicherheitsanalyse von Straßennetzen (ESN) – Erfahrungen mit den Verfahren der Abschnittsbildung. Straßenverkehrstechnik **52**(9), 515–527 (2008)
7. Thüringer Ministerium für Infrastruktur und Landwirtschaft (TMIL): Landesstraßenbedarfsplan 2030 – Entwurf. Entwurf, Erfurt (2018)
8. Elvik, R.: State-of-the-art approaches to road accident black spot management and safety analysis of road networks. TØI report 883/2007. Norwegian Centre for Transport Research, Institute of Transport Economics TØI, Oslo (2007)

Keeping the Autonomous Vehicles Accountable: Legal and Logic Analysis on Traffic Code

Dan M. Costescu(✉)

University "Politehnica" of Bucharest,
Spl. Independentei 313, 060042 Bucharest, Romania
dan.costescu@stud.trans.upb.ro

Abstract. The recent boost of the autonomous vehicles (AV) technology clearly outpaced the related regulating processes. Even for the already existing regulation like Traffic Code, opposable today to the human drivers, supposing to be maintained on medium term without important modifications, a major challenge pops-up. The content should by "translated" to a form enabling the engineers and authorities to keep also the autonomous vehicles accountable against the rules. The formalization methodology is consolidated, containing three major steps, practically implemented for the Overtaking as study case. Firstly, the Legal Analysis aims to eliminate the inherent redundancy of existing legal texts, clearly separate the responsibility of user and AV and finally to logically break them down in "*predicate precursors*". The following step, Logic Analysis was conceived as a bridge between legal and engineering aspects, shaping step by step the Overtaking maneuver by using tools like the Modal Logic Flow Chart (sequenced but still untimed events), Atomic Proposition Tables (multi-level hierarchy) and finally Linear Temporal Logic (LTL) Formulas timed over a Temporal Logic Diagram. At this point the goal of this paper has been achieved, the methodology delivering the appropriate preparedness for automation and for further improvements of expressivity using High Order Language (HOL) by performing Engineering Analyses for aspects not yet covered by literature.

Keywords: Autonomous vehicles · Legal analysis · Logic analysis · Traffic rules formalization

1 Introduction

Considering the complex multimodal scenarios featuring urban and inter-urban traffic, mixed categories of vehicle, at-grade intersections and less predictable human behavior even under clear traffic regulations, it is obvious that defining standards and achieving an airtight safety and security package to society is not as straightforward as expected for the autonomous vehicles (AV).

Although the terms safety and security are generally used as interchangeable, the literature makes a clear difference between them. According to Burns et al. [1] and Albrechtsen [2], security is protection against deliberate incidents and safety is

A. Varhelyi et al. (Eds.): VISZERO 2018, LNITI, pp. 21–33, 2020.
https://doi.org/10.1007/978-3-030-22375-5_3

protection against unintended incidents. A more general definition would define safety as addressing the ability of a system to avoid undesirable effects on its environment whilst the security refers the traits of an environment necessary to avoid undesirable effects on the system [3].

In spite of spectacular technology development, for a long period from now on the traffic will be still mixed, including both human operated and autonomous vehicles. A reasonable approach would be then to avoid any sudden and sharp change of traffic code that could break the principle of continuity and predictability of safety regulation. This is, for the next years the principles of keeping human drivers legally accountable will be maintained and accordingly that should equally apply to AVs. However, as the automation level of AV will increase gradually, the liability for already existing rules shall transfer from user to vehicle (manufacturer) according to the volume of automated functions (see Fig. 1) and on medium and long term it is anticipated that the traffic code will be completed with new technology specific rules. On both cases it is required a clear methodology of distributing responsibility between user and vehicle and then, for the rules applicable to AV to "translate" them from the legal idiom into a directly computable, free of ambiguities, language. This is the formalization of traffic rules, the topic present paper aims to contribute on.

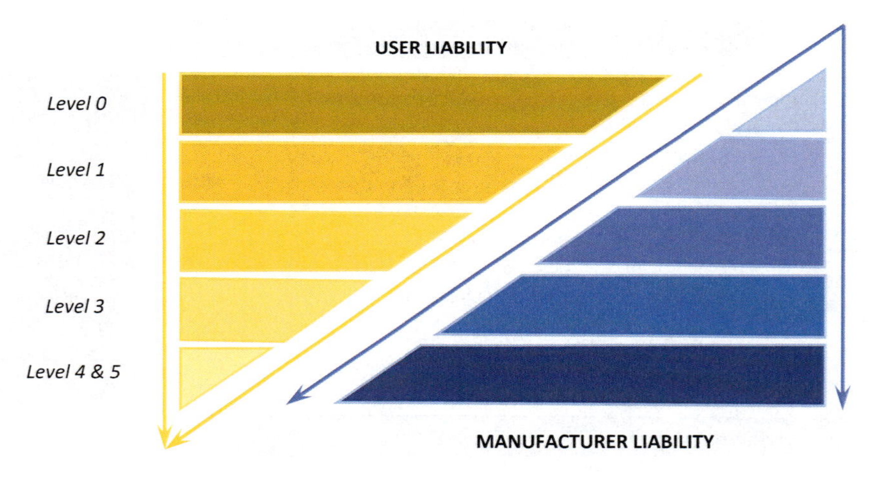

Fig. 1. The expected liability transfer from user to manufacture as automation level increases.

2 Formalization of Legal Acts

Literature assigns the work on the "The British Nationality Act" as the first major attempt to formalize a law into a logic program [4]. It has been proved that there are a lot of similarities between law formulation and computer languages, the former including since centuries before computer era, elements similar to programming structures, specifications, database characterization and query, integrity constraints and even knowledge representation specific to artificial intelligence [5].

2.1 Legal Analysis of Traffic Rules

Definition of Legal Scope, Associated Set and Subsets of Regulation. Theoretically, in respect with the traffic, a comprehensive set of regulation $\{R\}$ should be defined, that could be further broken apart in two different, preferably disjoint subsets:

$$\{R\} = \{User\} \cup \{AV\} \tag{1}$$

$$\{User\} \cap \{AV\} = \emptyset \tag{2}$$

where $\{R\}$ is the traffic regulation set including all traffic rules, $\{User\}$ – the traffic rules subset assigned exclusively to the user, $\{AV\}$ – the traffic rules subset assigned exclusively to the autonomous vehicle.

Equation (1) indicates that the $\{R\}$ is a gaps free regulation set, meaning there is no traffic rule which is not assigned at least to one of the two subsets, $\{User\}$ and $\{AV\}$.

Equation (2) indicates that $\{R\}$ is a regulation set free of shared responsibilities which is the ideal legal setup, where the responsibility for every situation is entirely assigned, either to the user or to the autonomous vehicle.

The two concepts introduced above bring clarity in separation of responsibility areas. Despite the fact the latter one is an idealized case, when the two defined subsets were disjoint, which is obviously not the case in reality, its consideration eliminates ambiguity and helps defining a third subset of rules, that needs special attention. This is the intersection of the first two subsets (Eq. 2), where the responsibility is shared or needs further clarification (e.g. cases when AV gives back control to human operator).

Practically, the above formulated problem, starts from an already existing set $\{R\}$ of traffic rules, approved at national level and expressed in natural language, more specifically, in terms of a legal dialect which is a specialized language but still a natural one. Assuming that the subset $\{AV\}$ was appropriately extracted from the set $\{R\}$ the next issue to be tackled is that the natural language is in the most cases not suitable for assessing whether the AV obeys the traffic rules or not. This is, the sometime vaguely formulated rules as "adapting the speed to the extent of avoiding any dangerous situation" makes almost impossible in this form, the design or monitoring in traffic of an AV, which is actually a matter of position, time, speed and acceleration [6]. Further steps are required, in order to make the present traffic codes directly usable for *engineering purposes*, namely defining technical specifications and for *enforcement purpose*, namely monitoring behavior in traffic and liability topics.

A combination of legal, logic and engineering analyses should be employed with the declared aim to first ***filter the unnecessary and ambiguous information***, then ***codify the traffic rules*** into predicates as 'overtaking pre-check', 'do overtake', 'adapt speed' or 'safe distance' and thereafter perform their ***concretization***, interpreting from an abstract form to a clear and practicable one for an automated approach.

Traffic Rules Codification. The principles of *propositional logic* [7] have been used to generate *formulas*, generally constituted of *atomic propositions (AP), as* statements without internal structure and *Boolean operators*. That technique is usually sufficient when preparing the "predicate precursors" for legal analysis. When propositional logic is not enough expressive to formalize the mathematics of engineering analysis, additional elements will be introduced, as *variables, arithmetic operators* and *relational operators*, and the resulted system of logic that can be interpreted on their basis is called a ***first-order logic*** or ***predicate logic*** [8]. For the cases where a finer distinction between a 'true' and 'false' value is needed, the ***modal logic*** is able to distinguish between statements that are *necessarily true* and those that are *possibly true.* For the specific processes of AVs, the ***temporal logic*** [9] as a form of modal logic proves to be more appropriate when time points should be referred, interpreting *necessarily* as *always* and *possibly* as *eventually*. Some of these formalisms will be used as *untimed* when time is referred only from the point of view of state ordering [10] whilst other formalisms will be employed further in the model because they allow declaring and using time values, mainly relative, but absolute as well [11].

Furthermore, the complexity of the autonomous vehicles and their processes asks to migrate from the first-order logic able to quantify only variables ranging over individuals, towards superiors levels as *second-order logic* that quantifies over sets, towards *third-order logic* that quantifies over sets of sets, and finally towards ***higher-order logic*** (HOL) that delivers quantification over sets that are arbitrarily deeply nested, being actually a union of 1st-, 2nd-, …nth-order logic [12, 13].

The Romanian traffic code [14] was used as a working document in its form including the *main body*, and the related *regulation procedures*. To avoid confusion the articles of the former will be referred as Art. only with its number whilst the articles of the later with its number preceded by an "R. Art.".

One of the most complex manoeuvers on road is *overtaking*, that remains actually the major challenge for traffic rules formalization and it is also selected as the case for the present study. A summary of the relevant legal provisions is presented in Table 1.

Legal Analysis of Traffic Rules: Overtaking Maneuver. A compared analysis has been performed between the paragraphs selected as relevant for the study case in Table 1. The aim was ***to eliminate the inherent regulating redundancy*** and to retain only a concise and coherent foundation for subsequent formalization. Text in italic indicates those topics being already treated by other paragraphs or simply not relevant to the context and therefore excluded from further processing, as redundant.

The next step is what we will call from now on "***the logical break down***" of the law text as per Fig. 2, the Art. 45 (1, 2, 3) being used as example. It can be seen the legal text suffers only minor modification if any, but it is broken down in sentences semantically significant (*predicate precursors*) separated by Boolean operators.

3 Logic Analysis

It was conceived to enable a smooth transition from Legal to Concretization stage. After processing of all the retained articles according to Table 1, these are assembled sequentially for Overtaking in what we call from now on a "*Modal Logic Flow Chart*" represented in Fig. 3, in terms of concepts as *necessarily true* and *possibly true* specific to *modal logic*.

45 (1)	Overtaking is the manoeuver through which:

1.1.1.	→	which one vehicle passes another vehicle that is on the same lane
		OR
1.1.2.	→	an obstacle that is on the same lane

being executed by:

1.2.1.1.	→	changing the running direction
		AND
1.2.1.2.	→	leaving the initial lane

OR

1.2.2.1.	→	changing the running direction
		AND
1.2.2.2.	→	leaving the initial row of vehicles

45(2)	The driver engaging in an overtaking must make sure

2.1.	→	the forerunning vehicle has not initiated a similar manoeuver
		AND
2.2.	→	the following vehicle has not initiated a similar manoeuver

45(3)	When during the overtaking the central line is crossed, the driver must make sure that:

3.1.	→	no vehicle approaches from the opposite direction
		AND
3.2.	→	there is enough space to complete the overtaking and
		THEN
3.3.	→	return on the initial lane

Fig. 2. Logical break down of Art. 45 (1-3).

Table 1. Relevant traffic provisions regarding overtaking.

Art. No	Topic	Content	Comments
Art. 45 *(1) (2) (3)*	Overtaking	"(1) Overtaking is the manoeuver through which one vehicle passes another vehicle or an obstacle that is on the same lane, being executed by changing the running direction and leaving the initial lane or row of vehicles (2) The driver engaging in an overtaking must make sure the forerunning and the following	(1) Overtaking definition (1) Types of overtaking (2) Making sure (3) Making sure

(continued)

Table 1. (*continued*)

Art. No	Topic	Content	Comments
		vehicle has not initiated a similar manoeuver (3) When during the overtaking the central line is crossed, the driver must make sure that no vehicle approaches from the opposite direction and there is enough space to complete the overtaking and return to the initial lane"	
Art. 54 *(1) (2)*	Manoeuvers	"Art. 54. (1) The driver of a vehicle changing the running direction, leaving or entering a raw of vehicles, changing the lane or (…) is obliged to timely signalize intention and make sure the traffic is not disturbed or the other road users are nor endangered through his manoeuvers (2) Signalization of any change in running direction shall be maintained during the entire duration of the manoeuver"	(1) Signaling obligation (1) Making sure *(dealt partially by 45(3),* except case of followers) (2) Signaling duration
R. Art. 116	Signalization	"*(1) The drivers shall signalize the changing of running direction, overtaking, stopping and putting into motion* (2) Intention of road users to change the running directions, to leave or enter a row of vehicles, to change the lane or (…) must be signalized by turning on the blinking lights with at least 50 m within urban areas and at least 100 m otherwise, before starting the manoeuver	*(1) Signaling obligation* *(d. by 54(1))* (2) Signaling distance
R. Art. 118	Overtaking	"The driver engaging in taking over is obliged to: *(a) Make sure the follower or forerunner did not signalize the intention of a similar manoeuvers and he can safely takeover without inducing any discomfort*	*(a) Making sure dealt also by 45(2)(3)* *(b) Signalizing dealt also by 54(1)* *(c) Lateral safe distance* *(d) Return obligation*

(*continued*)

Table 1. (*continued*)

Art. No	Topic	Content	Comments
		to the users running from opposite direction; *(b) Signalize overtaking intention;* (c) Keep during the manoeuver a sufficient lateral distance to the overtaken vehicle; (d) Return to initial lane or row of vehicles after making sure such a manoeuver is safe for all road users"	
R. Art. 120 (1) a ÷ k, (2)	Overtaking pre-check	A number of situations when the overtaking is forbidden are listed within this article. As it makes no difference and for the sake of concision, they will be further treated as a built in-set of conditions. However it worth to mention that the Pre-check set contains two subsets, one related to infrastructure R. Art. 120 {a, b, c, d, e, f, g, i} and one related to the other road users R. Art. 120 {j, k}. The later should be addressed together with Art. 45 (1), (2)	

A complete analysis of AV overtaking maneuver will be performed, to define the ***Atomic Propositions*** for the early stages of detecting, planning and making sure but also for the last stage of concretely performing the maneuver. Automation preparedness is improved by establishing a multilevel hierarchy. The logic formulas **for Linear Temporal Logic (LTL)** will be then expressed in terms of following: *Detecting-Need & Types-of-Overtaking, Pre-Check-Overtaking and Perform-Overtaking*. Finally, the ***Temporal Logic Diagram*** will be built, assigning time descriptions to the former untimed sequence of modal Logic Flow Chart and, by explicitly making use of the *temporal logic* concepts of *always true* and *eventually true*, allowing further development *in* ***High Order Language (HOL)***.

Modal Logic Flow Chart supplies sufficient information for further definition of atomic propositions as per Table 2, including both their interpretation and sequence. The three important phases, *Detecting Need & Type of Overtaking*, *Pre-Check Overtaking* and *Perform Overtaking* considered itself as level 1.

As described in logic interpretation, during the detection phase no maneuver is executed yet but the atomic propositions are assigned with the Boolean values in order

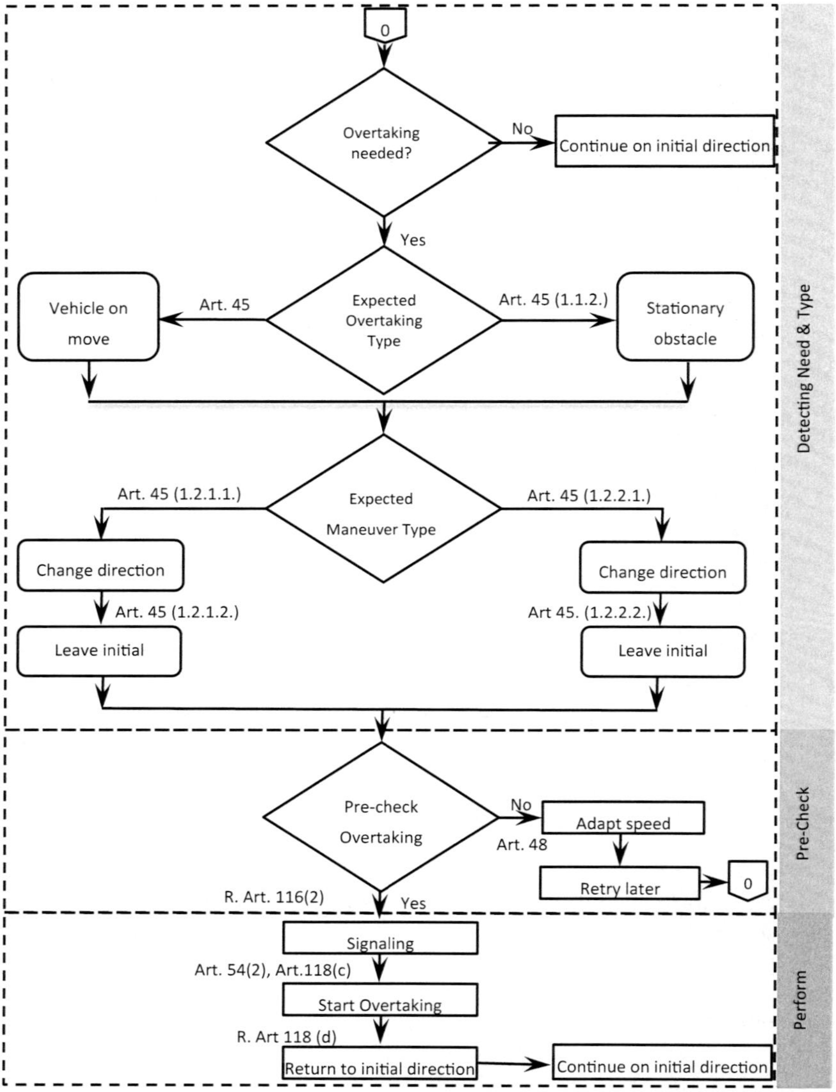

Fig. 3. Modal logic flow chart of overtaking.

to establish what type of overtaking if any, should be executed in a later phase. The assignments have following signification:

- for the level 2. AP *Overtaken-Needed?* value 1 and 0 simply indicates if AP is True or False,

Table 2. Atomic propositions, level and interpretations for *Detecting Need &Type of Overtaking*.

Level	Time	Atomic propositions (AP)	Logic interpretation
1.	$t_0 - t_3$	*Detecting-Need & Types-of-OT*	Determine if an OT is necessary and its type
2.	$t_0 - t_1$	*Overtaken-Needed?*	Determine whether an OT is necessary:{1, 0}
2.	$t_1 - t_2$	*Detect-Overtaken-Type*	Determine the type of overtaken entity:{m, n}. Coded as 1 when values m, n detected, 0 otherwise (before)
3.		*Vehicle-on-Move*	:=m, Coded as 1 if the entity to be overtaken is a vehicle on move, 0 otherwise
3.		*Stationary-Obstacle*	:=n, Coded as 1 if the entity to be overtaken is an stationary obstacle, 0 otherwise
2.	$t_2 - t_3$	*Detect-maneuver-Type*	Determine how to perform OT:{p, q, r} Coded as 1 when values p, q, r detected, 0 otherwise (before)
3.		*Change-Direction*	:=p, Coded as 1 if requested, 0 otherwise
3.		*Leave-Initial-Lane*	:=q, Coded as 1 if requested, 0 otherwise
3.		*Leave-Initial-Vehicle-Row*	:=r, Coded as 1 if requested, 0 otherwise

- for the level 2. APs, *Detect-Overtaken-Type* and *Detect-Maneuver-Type,* the value 1 means that the detection processes have been completed (the values m, n and respectively p, q, r were assigned with their Boolean values) and value 0 otherwise.

Any change of *Overtaken-Needed?* from 1 to 0 reset to 0 the APs *Detect-Overtaken-Type* and *Detect-Maneuver-Type* while any change of *Overtaken-Needed?* from 0 to 1 trigger restarting their assessment. If an overtaking is requested, at the final of detecting process, all three Level 2. APs should have value 1 (true) and the overtaking vector OT (o m n p q r) should be fully evaluated. E.g. an OT (1 0 1 1 0 1) means that an overtaken is requested, the obstacle is not a vehicle on move but stationary and the maneuver should be done by changing direction and not leaving the initial lane but only leaving the initial row of vehicles.

Formulation of traffic rules in Linear-Temporal-Language (LTL) based on atomic propositions completes the codification process, facilitates the logic automation and allow a formal model checking [9]. The resulted formulas for *Detecting Need & Type of Overtaking*, specific to above defined time intervals, are:

1. Overtaking identified as necessary:

$$\mathcal{F}_{(0-1)} = f(\text{Overtaken-Needed?}), \tag{3}$$

where $\mathcal{F}_{(0-1)}$ is the formula evaluated between the time points [t_0, t_1], $f(AP)$ – the function assessing the atomic proposition AP always as true.

2. Overtaking is the maneuver through which one vehicle passes another vehicle or an obstacle that is on the same lane -Art 45, (1.1.1) & (1.1.2):

$$\mathcal{F}_{(1-2)} = f(\text{Detect-Overtaken-Type}{\rightarrow}\text{Vehicle-on-Move} \oplus \text{Stationary-Obstacle}) \quad (4)$$

3. Overtaking is the maneuver (..) being executed by changing the running direction and leaving the initial lane or row of vehicles - Art 45 (1.1.1.1) & (1.1.1.2) & (1.1.2.1) & (1.1.2.2):

$$\mathcal{F}_{(2-3)} = f(\text{Detect-Maneuver-Type}{\rightarrow} ((\text{Change-Direction} \wedge \text{Leave-Initial-Lane}) \oplus (\textit{Change-Direction} \wedge \textit{Leave-Initial-Vehicle-Row}) \quad (5)$$

4. When overtaking identified as necessary that is the maneuver through which one vehicle passes another vehicle or an obstacle that is on the same lane, being executed by changing the running direction and leaving the initial lane or row of vehicles – Art 45 (1) & (2):

$$\mathcal{F}_{(0-3)} = f(\text{Detecting-Need \& Types-of-Overtaking}{\rightarrow}\text{Overtaken-Needed? }{\rightarrow}\text{Detect-Overtaken-Type}{\rightarrow}\text{Detect-Maneuver-Type}) \quad (6)$$

For the brevity the similar formulation of *Pre-Check of Overtaking* and *the Perform-Overtaking* are not included in the present work. The final logic formulas of them are:

1. Overtaking is safe as both conditions related to infrastructure and other users are fulfilled R. Art. 120 (1) (2):

$$\mathcal{F}_{(3-5)} = f(\textbf{Pre-Check-Overtaking}{\rightarrow} (\text{Pre-Check-Infrastructure} \wedge \text{Pre-Check-Road-Users}) \quad (7)$$

2. Perform overtaking according to Art 45 (1), (2), (3), R. Art. 116 (2), R. Art. 118, c - d:

$$\mathcal{F}_{(5-9)} = f(\textbf{Perform-Overtaking}{\rightarrow}\text{Signaling}{\rightarrow}\text{Start-Overtaking}{\rightarrow}\text{Advance-Overtaking}{\rightarrow}\text{Complete-Overtaking}) \quad (8)$$

Finally, the logic formulation of Overtaking maneuver in Linear Temporal Language is expressed as follows:

$$\mathcal{F}_{(0-9)} = f(\textbf{Overtaking}\rightarrow\textbf{Detecting-Need \& Types-of-Overtaking} \rightarrow\textbf{Pre-Check-Overtaking}\rightarrow\textbf{Perform-Overtaking}) \tag{9}$$

and the Overtaking is now fully coded from the moment t_0 when the need is identified, till t_9 when the maneuver is completed. The processes' sequence related to the above described atomic propositions and logic formulas is graphically presented in Fig. 4.

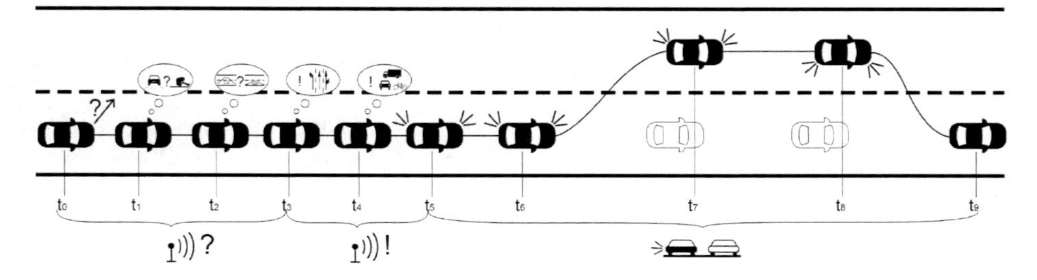

Fig. 4. Temporal logic diagram.

The three major phases of overtaking were logically coded over their validity time intervals: *Detecting-Need & Type-of-Overtaking* ($t_0 \div t_3$), *Pre-Check-Overtaking* ($t_3 \div t_5$) and *Perform-Overtaking* ($t_5 \div t_9$) and they are now fully prepared for automation.

4 Conclusions and Future Work

The Traffic Code Violating Monitor was proposed in order to keep the autonomous vehicles accountable against the traffic rules. Usually these rules are consolidated in National Traffic Codes and approved as laws by the authorities, being therefore expressed using legal specific terms, syntax and semantic which improve the precision comparing with natural language but not enough to allow an immediate automation.

A *Legal Analysis*, as a first step to resolve ambiguity is required but a concrete approach is still missing by now in recent literature. The author proposed separation of the traffic code in two major subsets of rules, concerning the responsibility of the user and AV, respectively. Two concepts are newly introduced, first being the *Gaps Free Set of Regulation,* meaning there is no traffic rule which is not assigned at least to one of the two subsets, which is important to consider for innovative and disruptive technologies. The second concept is that of a *Set of Regulation Free of Shared Regulation* meaning that the responsibility for every situation is entirely assigned, either to the user or to the autonomous vehicle. That is an idealized case, when the two defined subsets were disjoint, which is obviously not the case in reality, but its consideration helps defining a third rules subset, the intersection of the first two major subsets, that needs special attention.

Mandatory steps of legal analysis were defined and then implemented for the Overtaking case study, namely the *Elimination of inherent regulation redundancy* and *Logical break down of legal text*.

The *Logic Analysis* was conceived as the transition from Legal to Engineering Analysis and it consist of *Modal Logic Flow Chart* that assemblies sequentially the elements of the modelled maneuver, making use explicitly of *modal logic* concepts of *necessarily true* and *possibly true*.

Consequently, a complete analysis of AV's overtaking has been performed, generating for the first time the *Atomic Propositions* not only for the last stage of concretely performing the maneuver but also for the early stages of detecting, planning and making sure. For helping automation, a multilevel hierarchy has been proposed within the *Atomic Proposition Table.*

The logic formulas for Linear Temporal Logic (LTL) have been expressed in terms of atomic propositions for *Detecting-Need & Types-of-Overtaking, Pre-Check-Overtaking and Perform-Overtaking*. Finally, the *Temporal Logic Diagram* is constructed assigning time descriptions to the former untimed sequence of Modal Logic Flow Chart and allowing further development in High Order Language (HOL) by making use explicitly of the *temporal logic* concepts of *always true* and *eventually true.*

After facilitating a *better automation* through the logic analysis of overtaking rule, the *expressivity* should be enabled by *Engineering Concretization* in a future work. The author proposed to develop for the first time the overtaking model for three vehicles where at least the overtaken vehicle is AV, as a basis for Traffic Code Violating Monitor.

References

1. Burns, A., McDermid, J., Dobson, J.: On the meaning of safety and security. Comput. J. **35**(1), 3–15 (1992)
2. Albrechtsen, E.: A generic comparison of industrial safety and information security. In the PhD course "Risk and Vulnerability", Norwegian University of Science and Technology (2002)
3. Line, M.B., Nordland, O., Røstad, L., Tøndel, I.A.: Safety vs security? In: Proceedings of the 8th International Conference on Probabilistic Safety Assessment & Management. ASME Press, New York (2006)
4. Sergot, M.J., Sadri, F., Kowalski, R.A., Kriwaczek, F., Hammond, P., Cory, H.T.: The British Nationality Act as a logic program. Commun. ACM **29**(5), 370–386 (1986)
5. Kowalski, R.A.: Legislation as logic programs. In: Bankowski, Z., White, I., Hahn, U. (eds.) Informatics and the Foundations of Legal Reasoning. Law and Philosophy Library, vol. 21, pp. 325–356. Springer, Dordrecht (1995)
6. Rizaldi, A., Althoff, M.: Formalizing traffic rules for accountability of autonomous vehicles. In: IEEE 18th International Conference on Intelligent Transportation Systems (ITSC), pp. 1658–1665 (2015)
7. Ben Ari, M.: Mathematical Logic for Computers. Springer, London (2012)
8. Basin, D.A., Klaedtke, F., Mueller, S., Zalinescu, E.: Monitoring metric first-order temporal properties. J. ACM **62**(2), 15 (2015)
9. Clarke, E.M., Orna, G., Peled, D.: Model Checking. MIT Press, Cambridge (1999)

10. Pnueli, A.: The temporal logic of programs. In: Proceedings of the 18th Annual Symposium on Foundations of Computer Science. IEEE Computer Society (1977)
11. Raskin, J.F.: Logics, Automata and Classical Theories for Deciding Real Time. Facultés universitaires Notre-Dame de la Paix, Namur (1999)
12. Shapiro, S.: Classical logic II: higher order logic. In: Goble, L. (ed.) The Blackwell Guide to Philosophical Logic, pp. 33–54. Blackwell, Oxford (2001)
13. Benzmueller, C., Miller, D.: Automation of higher-order logic. In: Gabbay, D.M., Siekmann, J., Woods, J. (eds.) Handbook of the History of Logic. Computational Logic, vol. 9. Elsevier, North Holland (2014)
14. Ursuta, M.: Codul Rutier. Universul Juridic, Romania (Traffic Code of Romania) (2016)

Causes of Road Accidents with Fatalities and Heavy Injuries in Latvia

Juris Kreicbergs[1](✉), Oskars Irbitis[1], and Janis Kalnins[2]

[1] Riga Technical University, 1 Kalku Str., LV1658 Riga, Latvia
Juris.Kreicbergs@rtu.lv
[2] Ministry of Transport, 3 Gogola Str., LV1743 Riga, Latvia

Abstract. From all traffic accidents the accidents with fatalities and with heavy injuries are of the most concern of traffic safety authorities. Although a reasonable decline rate in road fatalities has been achieved, Latvia still is among the countries in European Union with the highest number of road deaths per number of inhabitants. Traffic safety research in Latvia for a range of years has been constrained by limited access to road accident documents by researchers due to legal reasons. The current research lead by Ministry of Transport coordinated researchers from State Police and State Forensic Science Bureau who had full access to road traffic accident documentation, Road Traffic Safety Directorate managing the statistics data, Latvian State Roads having expertise in road infrastructure safety assessment and Riga Technical University having skills in traffic safety research to analyze all road accidents with fatalities and heavy injuries in Riga region in 2016. Along with descriptive statistics data the research revealed the causes of the high percentage of vulnerable road users among the fatalities and heavily injured, problems with pedestrian crossings and other road infrastructure, certain road user behavioral problems and shortcomings in regular data collection for road traffic safety analysis. The findings will help to fulfill the goals of reducing the numbers of fatalities and heavily injured on roads and have clarified the course for future traffic safety research in Latvia.

Keywords: Road fatalities · Heavy injuries · Latvian roads

1 Introduction

1.1 A Subsection Sample

Road traffic accidents cause approximately 1.3 million fatalities worldwide, more than half are among vulnerable road users: pedestrians, cyclists, and motorcyclists [1]. More than 25 thousand lost their lives in EU in 2017 [2], pedestrians accounted for 21% of all victims, 25% were two-wheelers (14% motorcyclists, 8% cyclists and 3% moped riders) [3]. In Latvia among road fatalities in 2017 there were 37.5% pedestrians and 13.2% two-wheelers (0.7% motorcyclists, 8.1% cyclists and 4.4% moped riders) [4] which is closer to worldwide than to EU statistics. Although Latvia has the second biggest decrease of road deaths per million inhabitants from 2001 to 2017 [5], in absolute numbers with 70 deaths/mln Latvia is still well below the EU average of 50 and far below the Norwegian minimum of 20. Traffic safety research in Latvia for years

A. Varhelyi et al. (Eds.): VISZERO 2018, LNITI, pp. 34–40, 2020.
https://doi.org/10.1007/978-3-030-22375-5_4

mainly emphasized the analysis of accident statistics by Road Traffic Safety Directorate while exploration of road accident causes has been largely constrained by limited access to road accident documents by researchers due to legal reasons. Therefore the current research initiated and coordinated by Ministry of Transport, involving researchers from State Police and State Forensic Science Bureau who had access to road traffic accident documentation, gave more insight into tragedies on Latvian Roads.

There is no single factor that drives road casualties [6]. The causes of road accidents are described by many governmental and non-governmental organizations, insurance and other companies, researchers, bloggers and enthusiasts. Most blame drivers (distracted, reckless, drunk driving, speeding), weather conditions, more rare mechanical failure, medical condition or animals on road. Some take a wider scope, looking at economic conditions, behavioral problems in society and governmental policies. Knowing the general causes of road deaths but understanding that traffic safety figures are not uniform across countries, when it comes to policy implementation, there is a willingness to get confirmation if and how these causes have worked for a particular region looking through every road accident with heavy casualties and trying to assess what could be done to prevent similar accidents.

2 Materials and Methods

The goal was to analyze as many recent heavy road accidents as possible to find patterns leading to loosing lives and injuring people on Latvian roads. The research was done in 2017 and had to provide with interim results by the end of the year. Since the accident documentation is stored on different locations, the sample size of the heavy road transport accidents investigated was limited by all accidents with fatalities and heavy injuries occurred in Riga region in 2016.

To get access to comprehensive accident data the involvement of researchers from the State Police was essential. Since the police have their analytical staff mostly in Riga and the accident documentation is stored in regional stations, to avoid additional travel, the accidents in Riga region were sampled only. The police researchers supplied with the accident timing, place, including the number of reoccurrences on the same spot within three years, data on accident type, people involved, including age, gender, citizenship, role at the accident, injuries, fatalities, usage of seat belts, helmets reflectors, being or not under influence of alcohol or other substances. The police protocols also supplied vehicle data, including type, model and make, year of production, vehicle annual inspection results, tire type and condition, weather conditions, road surface type and condition, lighting, road accident schematics and photos taken on the accident sites. Knowing that for several reasons the police reports may underreport the accident factors [7], these data have been verified against the databases and road audits performed by Road Transport Safety Directorate and a Road infrastructure expert from Latvian State Roads visited every accident site except one located in private territory and evaluated the site infrastructure against Latvian standards.

Experts form State Forensic Science Bureau checked if there has been a previous court investigation, analyzed the mechanics and causes of the accidents including the primary and secondary accidents, the best estimate of vehicle speed and possibility of speeding over the allowed limit, the accident influencing factors and their causal relation with accident occurrence and accident consequences, the technical possibility of accident prevention by the people in the accident.

All experts from the organizations mentioned submitted their raw data and reports to Transport Ministry where a summary report has been produced [8]. The current paper describes the findings from the raw data available and from the interim and final reports of the research.

3 Results and Analysis

3.1 Accident Statistics

Total numbers for accidents and casualties in Latvia and the research sample size are given in Table 1. The sample size represents above 60% of accidents and accidents with casualties wile just slightly over 30% fatalities or heavily injured suggesting that accidents are different from the ones in the rest of Latvia. The Riga region contains the biggest urban area in Latvia therefore 65% of the accidents analyzed occurred in urban areas having lower speeds and therefore less severe consequences.

Table 1. Accident and casualties' population and sample size.

Parameter	Data for Latvia 2016	Riga region data	Percentage
Number of accidents	19,527	12,534	64
Accidents with casualties	3,792	2,350	62
Heavily injured	528	175	33
Heavily injured investigated		40	23
Fatalities	158	48	30
Accidents investigated		84	

From all accidents examined two thirds involve vulnerable road users (59% involve pedestrians, 7% cyclists) and 65% have occurred in urban areas. Since this paper is prepared for Vision Zero for Sustainable Road Safety in Baltic Sea Region conference section Urban planning and vulnerable road users, even if the selected sample may not fully represent the situation in Latvia, it well fits for the analysis of road accident casualties in urban and close to urban areas. This paper reports mostly about the findings of the research on vulnerable road users and policies from the Vision Zero [9] perspective.

3.2 Vulnerable User Casualties

The accidents investigated involved 48 fatalities, 24 of them were pedestrians, 15 car and truck drivers, 3 passengers, motorcyclists and cyclists each. From the total 48 fatalities in the accidents investigated 24 are pedestrians and 3 cyclists. Having this high percentage of pedestrians among fatalities requires more in-depth analysis. The location of pedestrian fatalities is shown on Fig. 1.

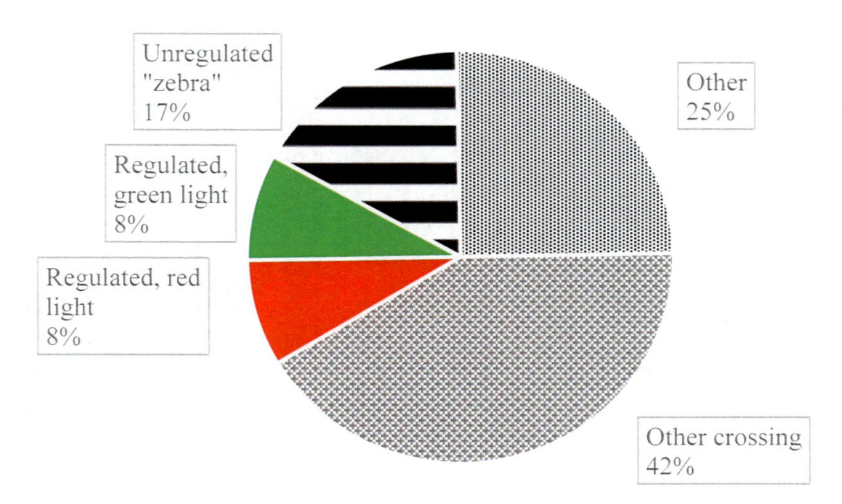

Fig. 1. Accident sites with pedestrian fatalities.

One fourth of all pedestrian fatalities have occurred where pedestrians had priority over other road users – 17% or 4 fatalities on non-regulated zebra crossings, 8% on regulated crossings with green light on. The research data on pedestrian even showed 48% of pedestrian serious injuries have occurred on unregulated zebra crossings but since in contradiction to fatalities, not all accidents with heavy pedestrian injuries have been analyzed this high percentage may be influenced by accident cases selection. The common feature for four fatalities on zebra crossings are wet road surface and sober pedestrians, other characteristics vary. Two occurred in day time, two at night with special lightning on, two crossings have been recognized as not corresponding to standards – one is put on too broad street, one has restricted approach visibility. For 12 cases of heavy injuries on zebra crossings in 5 cases the crossings are on too wide street, only 7 are on wet road and 5 at dark time of the day, 4 pedestrians have been under influence of alcohol, one driver was blinded by sun, one pedestrian has had earphones with loud music restricting adequate situation awareness. The situation seems similar to Poland [10, 11], where due to growing number of high risk pedestrian crossings (in Poland there is even higher speed allowed at zebra crossing sites) the percentage of casualties on zebra crossings is much too high.

In the accidents investigated 12 pedestrians or half of all pedestrians killed have crossed the road while they had to wait for free road or allowing traffic signal, 10 of them there under influence of alcohol. The problem is shown on Fig. 2, where the number of fatalities and heavy injured pedestrians are split according to their age. Alcohol usage has contributed to 47% of pedestrian casualties. In age groups from 20 to 70 years, 59%, in age groups from 20 to 40 years, above 85% of pedestrian casualties are alcohol related.

From three cyclists heavily injured and three killed two were under influence of alcohol, both resulted in fatalities. The third one and two of heavily injured resulted after vehicle not giving way after driving direction change showing the existing problem of not getting used to increasing cycling traffic.

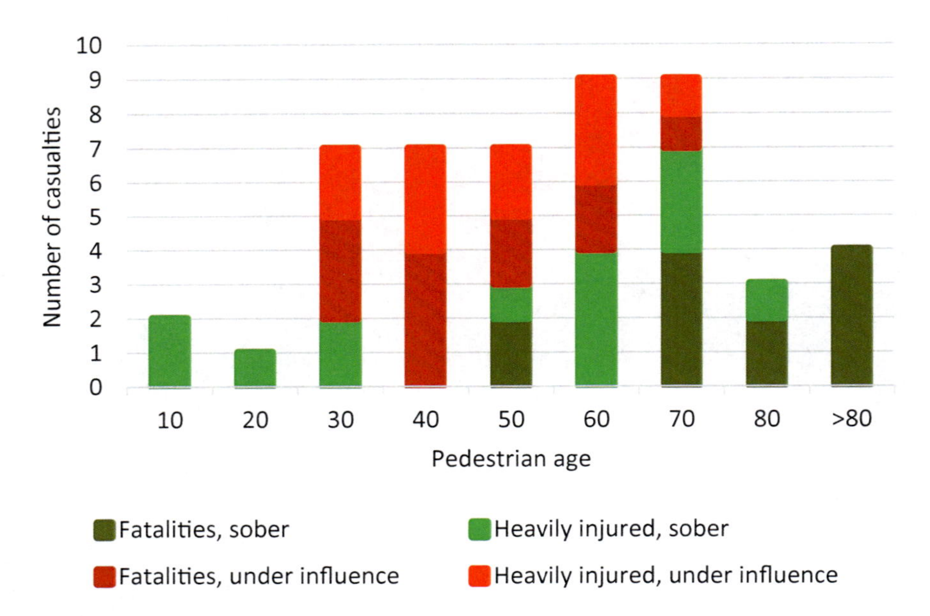

Fig. 2. Number of casualties according to pedestrian age and alcohol usage.

From three motorcyclists two were under the influence of alcohol, one was proved to be speeding, in two cases drivers did not notice the motorcycle.

In contrast no car or truck driver under influence has been registered in the accidents with vulnerable user casualties. Although drivers' alcohol usage in Latvia has significantly reduced [4], in the 84 accidents investigated from 96 drivers 9 have been under influence causing deaths of 6 drivers and two passengers.

3.3 Driver and Passenger Casualties

For 15 drivers killed in the accidents there is no single pattern, but mostly widely known causes of accidents. For two vehicles speeding was confirmed by accident reconstruction and for three others speeding could be involved to facilitate the severity of the accidents including two rollovers. Three drivers under influence of alcohol have aided their own deaths by running off the road in two cases and into opposite traffic lane in one case. A single death only was facilitated by snowy weather and correspondingly inappropriate speed, one fatality could be caused by fatigue and speeding combination, one death was initiated by driver by not securing a truck against movement while connecting a trailer, two may have had health problems and in two cases it was not possible to reconstruct the cause of running into opposite direction due to weak accident documentation in one case and not available accident documentation due to legal reasons in the other case.

There was no specific driver or vehicle age pattern – drivers almost of all age groups have suffered and both quite new and old cars have been involved. Two thirds of the vehicles were older than 12 years giving a chance that part of drivers can be saved in the future due to increasing passive safety of the vehicles.

Deaths of three passengers in separate accidents have been facilitated by two drivers under influence of alcohol, one of them speeding at double of allowed speed and one by non-appropriate tire usage on slippery road in combination with no seat belt usage.

3.4 Accident Analysis from the Vision Zero Perspective

From the Vision Zero perspective it is especially important to evaluate how road traffic system designers could have contributed to avoidance of traffic accidents analyzed, therefore for every accident analysis the State Forensic Science Bureau experts added their evaluation of what could be done to prevent the accident.

The expert from Latvian State Roads visited every accident site and on 83 sites found 33 cases where the traffic organization should be improved to match the existing standards. In 24 cases the traffic organization deficiencies most probably did not provoke the accidents analyzed, but were documented for site rebuilding. The problems were found on 13 zebra crossings, with 10 traffic signs, with 7 traffic lights assemblies, 2 road markings and one speed bump.

In nine cases it was concluded that improper zebra crossings may have directly contributed to road traffic accidents, in two cases on state roads and seven cases in urban areas. The major non-compliance with standards was building a zebra crossing across more than one traffic line in the driving direction. Some zebra crossings have been built close to a place where the main road changes its direction, on one crossing a pedestrian could appear from a cover of very close building. No pedestrian crossing has been found with allowed speed above 50 km per hour which is set as a maximum in the standard, but from Vision Zero approach and as it was confirmed by the unfortunate fatalities, even this speed may be too high for avoidance of casualties.

In many cases in the research two tendencies that contradict to Vision Zero approach is that for each accident where it is possible the road user actions are attributed to the cause of casualties and traffic organization is mostly compared against existing standards without analysis of how the standards may be improved. At the same time the research suggested more than 70 changes in traffic organization at the 84 accident sites and several actions at the traffic policies level. Above twenty improvements have to be done for the pedestrian infrastructure on the accident sites and the high number of non-corresponding infrastructure elements suggested that there has to be made a major revision of pedestrian crossings at least for the correspondence to existing standards. Almost twenty changes in traffic organization on the accident sites have been suggested but in four cases only the maximum permitted speed was suggested to be reduced.

Numerous suggestions of the research have been targeted to the behavior change of the road users, infrastructure improvement, assurance of vehicle technical condition and to the activities on traffic accident sites.

4 Conclusions

The road casualties research coordinated by Latvian Ministry of Transport with researchers from various organizations gave chance to analyze the causes of fatalities on Latvian roads in a way that is not possible by academic researchers due to accident documentation legal availability. The challenge of the research involving that many researcher groups working on their own set of information is that every group works with methods and goals they have been used to in their routine operation.

Vulnerable road user safety is an increasing concern of road traffic policy makers. There existing pedestrian crosswalks have to be re-examined, and the approach towards safe pedestrian mobility needs to be revised. Besides the reasonably achieved and on-going diminishing in drivers' alcohol usage, attention has to be emphasized towards pedestrian under influence fatalities on the roads.

There are diverse reasons of drivers' and passengers' fatalities, the contribution of speeding and alcohol usage is still among the main causes.

The research performed in Riga region with lower fatalities to number of accidents ratio may not fully characterize the situation in Latvia therefore it may be substantial to extend the research to other Latvian regions.

Acknowledgments. The research was financed by legislated deductions from the premiums of the compulsory civil liability insurance of the owners of motor vehicles. Special thanks are also to the staff of State Police, State Forensic Science Bureau and Road Traffic Safety Directorate who worked through the accident documents, Latvian State Roads who visited and analyzed every accident site and Ministry of Transport for coordinating the research.

References

1. World Health Organization Homepage. https://www.who.int/news-room/fact-sheets/detail/road-traffic-injuries. Last accessed 13 Jan 2019
2. Adminaite, D., et al.: European transport safety council, ranking EU progress on road safety. 12th Road Safety Performance Index Report, Brussels (2018)
3. European Commission - Fact Sheet: 2017 road safety statistics: what is behind the figures? Brussels (2018)
4. CSDD Homepage. https://www.csdd.lv/en/road-accidents/the-road-traffic-safety-statistics. Last accessed 21 Nov 2016
5. ETSC Homepage. https://etsc.eu/euroadsafetydata/. Last accessed 21 Nov 2016
6. Reported road casualties in Great Britain: 2017 Annual Report. Department of Transport, London (2018)
7. Rolison, J., et al.: What are the factors that contribute to road accidents? An assessment of law enforcement views, ordinary drivers' opinions, and road accident records. Accid. Anal. Prev. **115**, 11–24 (2018)
8. Kalnins, J., et al.: Comprehensive Research on Road Traffic Safety Influencing Risk Factors in Riga Region. Ministry of Transport, Latvia (2018). (in Latvian)
9. Kristianssen, A., et al.: Swedish Vision Zero policies for safety – a comparative policy content analysis. Saf. Sci. **103**, 260–269 (2018)
10. Pawlowski, W., et al.: Risk indicators for road accident in Poland for the period 2004–2017. Cent. Eur. J. Public Health **26**(3), 195–198 (2018)
11. Olszewski, P., et al.: Pedestrian fatality risk in accidents at unsignalized zebra crosswalks in Poland. Accid. Anal. Prev. **84**, 83–91 (2015)

Training of Road Safety Auditors in Germany

Julius Uhlmann(✉), Johannes Vogel, and Uwe Plank-Wiedenbeck

Bauhaus-Universität Weimar, Marienstraße 13D, 99423 Weimar, Germany
{julius.uhlmann, johannes.vogel, uwe.plank-wiedenbeck}@uni-weimar.de

Abstract. According to the Directive 2008/96/EC of the European Parliament and Council: "Member States shall ensure that road safety audits are carried out for all infrastructure projects". While the directive also lists certain criteria for training, the precise training curricula are given into the responsibility of the member states. As a result, the curricula and the organizations offering the training differ between the member states. While in some states, the responsible ministry organizes the training, others rely on private institutions or professional organizations.

This contribution discusses the advantages and disadvantages of the German model of the training of Road Safety Auditors where the execution of training courses for auditors is carried out by a cooperation of German Universities. An expert committee at the Road and Transportation Research Association develops guidelines for the training curricula.

Keywords: Road safety audit

1 Introduction

Road Safety Audits (RSA) are a mean to increase the road safety by letting special trained expert conduct analyses, called audits, of network elements (roads, intersections, etc.). Those audits can happen either in the planning process or when the network element is already in operation.

Since 2008, the member states of the European Union are required by the Directive 2008/96/EC to "ensure that road safety audits are carried out for all infrastructure projects" [1]. The competence for the training of the auditors is given into the responsibilities of the member states leading to a variation in points of training duration, responsible organization and curricula: While in some states professional or private organizations organize the training, in other ministries or universities do so. Also, the duration of the training courses varies between one day and several weeks [2].

In Germany, the training of auditors is carried out by a cooperation of German universities from the field of Transportation or Civil Engineering and lasts 11 days of seminars a plus ~ 100 h in additional exercises.

A. Varhelyi et al. (Eds.): VISZERO 2018, LNITI, pp. 41–46, 2020.
https://doi.org/10.1007/978-3-030-22375-5_5

2 Regulations for the Training of Road Safety Auditors

2.1 European

According to article 9 of Directive 2008/96/EC "Member States shall ensure that where road safety auditors carry out functions under this Directive, they undergo an initial training resulting in the award of a certificate of competence, and take part in periodic further training courses" [1]. More precise regulations, especially regarding the duration and contents of the training, are not set in the directive and given into the responsibility of the member states.

2.2 National

The German Recommendations for Safety Audits of Roads *(Empfehlungen für das Sicherheitsaudit von Straßen (ESAS))* require an auditor to have higher education or equivalent qualification in the field and several years of experience in the design of roads or in road safety. In addition, further competences should be acquired through additional training [3]. Precise contents and curricula for the training are not set.

In the last years, the ESAS was revised by an expert committee committee at the Road and Transportation Research Association *(Forschungsgesellschaft für Straßen- und Verkehrswesen (FGSV))* and the new version will be published soon. In the course of the revision the character of the document was upgraded from an R2 recomendation to an R1 regulation and accordingly named Regulations for Safety Audit of Roads (*Richtlinien für das Sicherheitsaudit von Straßen (RSAS))* [4].

In the Circular Letter for Road Construction No. 26/2010 (*Allgemeines Rundschreiben Straßenbau Nr. 26/2010)* of the German Ministry of Transportation, Construction and Urban Development from November 11, 2010, the ESAS is set as the standard for RSA in Germany. The training of auditors should take place according to the rules set in the Guidelines for Training and Certification for Road Safety Auditors (*Merkblatt für die Ausbildung und Zertifizierung der Sicherheitsauditoren von Straßen (MAZS))* [5].

The MAZS describes the requirements for auditors, the curricula of the training courses and the certification and recertification process of auditors [6]. The revision process of the MAZS is currently ongoing [7].

3 Training Curricula in Germany

3.1 General

The training curricula published in the MAZS are developed by an expert committee at the FGSV and give information on the course goals and contents and provide a guideline for the teaching methods.

The training is structured in modules (compare Fig. 1). The two foundation modules M1 and M2 are mandatory for all auditors. The auditor has to choose two to five of the thematic modules for different road types: motorways, rural roads, cross-town roads, urban main roads, and residential roads. The modules itself consist of lectures, exercises, assignments and site visits. An auditor is only certified to audit the types of roads he received training for. Although the MAZS lists some combination of thematic modules as examples, the combination of modules can be chosen freely.

The training is concluded with an exam in which the candidate has to audit a set of road projects according to the thematic modules the candidate received training for [6].

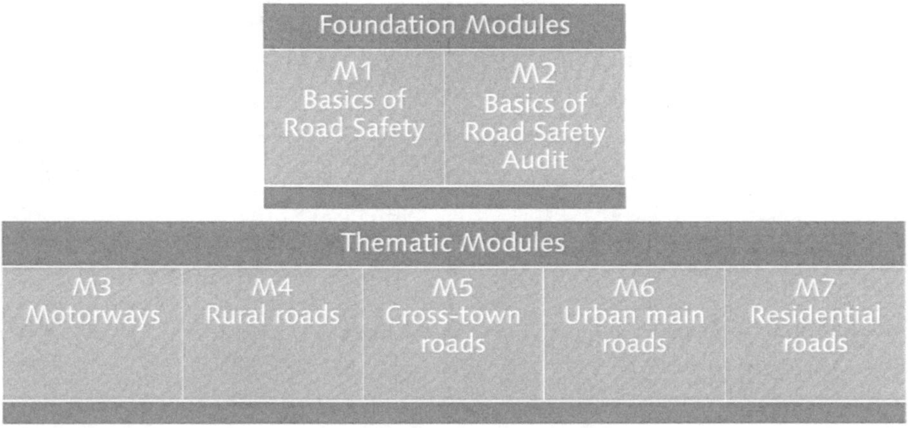

Fig. 1. Structure of the training curriculum of road safety auditors in Germany according to [6].

3.2 Example for Application

Figure 2 shows the structure of a training course for auditors for motorways, rural roads and cross-town roads as it is offered at the Bauhaus-University in Weimar. It can be seen that the workload for attendance (11 days) and for homework ($\sim$100 h) is almost equal. This way a high level of practical training during the course is achieved. The planning documents used in the training are partly submitted by the participants, hence a connection to current planning is given. The homework assignments have to be submitted to the university and will be corrected. The total duration of a training course is circa six months.

The seminars taking place in Weimar consist of lectures, short exercises, site visits and the discussion of the homework assignments. Approximately 36% of the seminar time accounts for lectures, 29% for the discussion of the homework assignments, 20% for exercises and their discussion and 15% for site visits and their discussion [8].

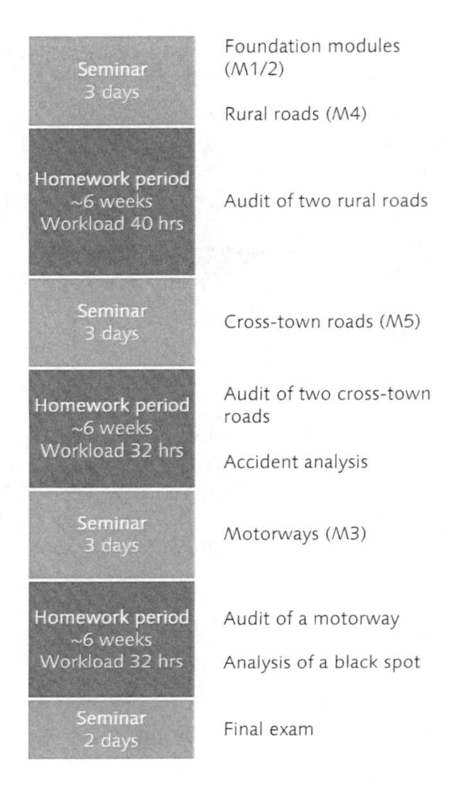

Fig. 2. Timeline of a training course for auditors as in operation in Weimar according to [9].

3.3 Further Training

The initial certification is valid for three years. To ensure a continuous and further training, auditors are required to during the three years:

- conduct on average at least one audit per year. If the auditor does not have the possibility to conduct real audits, he might conduct training audits with materials provided by the training institution [6];
- participate in at least two seminars regarding road safety. An example of those seminars are the symposia for road safety organized by the Road and Transportation Research Association at the universities in Wuppertal and Weimar [6]. The symposia consist of presentations on topic of road safety and workshops dealing with examples from the auditor's practice.

4 Discussion

With its high numbers of hours for both seminars and homework, the training of Auditors in Germany has a high standard. The homework training with real planning documents and site visits during the seminar provides good preparation for the tasks of an Auditor. As the candidates need to choose between specialties in the thematic modules, a strong thematic focus on the chosen field results.

The fact that universities from the field of transportation or civil engineering offer the training courses and are also included in the further development of the guidelines and training curricula, ensures the inclusion of new insights from road safety research in the training of auditors. As the seminars are usually visited by candidates from different road administrations and planning offices, they also offer a possibility for an exchange on professional topics.

The high level of training requires many resources at the training facilities for the preparation and realization of the seminars and the preparation and correction of the homework tasks.

The high number of required hours for the training courses, equaling one full month of work, can make the decision hard for an employer to allow a candidate to leave for the training. Internal training courses with instructors from the universities visiting the road administrations aim to reduce this problem.

The currently ongoing revision of the MAZS, conducted by a working team at the FGSV, tries to incorporate findings from the more than 15 years of training course expertise into the new recommendations for the training curricula. In discussion are changes in the structure and contents of modules to better train candidates regarding psychological aspects, accessibility, traffic lights, and auditing in teams. The new contents of the RSAS, especially the audit of roads in operation, should also be incorporated in the training curricula.

5 Conclusions and Outlook

The training of Road Safety Auditors in Germany is characterized by a high number of training hours and is based on a combination of theoretical seminars and practical assignments. In the years after the implementation, the general approach has proved itself and the experience gained in the training courses was used to further improve the training at a local level. The Guidelines for Training and Certification for Road Safety Auditors are currently in the process of updating through an expert committee at the FGSV and the new version is expected for 2021, incorporating findings from the training practice [7].

References

1. Directive 2008/96/Ec of The European Parliament and of the Council of 19 November 2008 on road infrastructure safety management (2008)

2. Vardaki, S., Dragomanovits, A., Gaitanidou, E., et al.: Development of a training course for road safety auditors in Greece. In: Transportation Research Board (ed.) TRB 95th Annual Meeting Compendium of Papers. (2016)
3. Forschungsgesellschaft für Straßen- und Verkehrswesen: Empfehlungen für das Sicherheitsaudit von Straßen (ESAS), Köln (2002)
4. Forschungsgesellschaft für Straßen- und Verkehrswesen: Arbeitsausschuss 2.7 Sicherheitsaudit von Straßen. https://www.fgsv.de/gremien/strassenentwurf/sicherheitsaudit.html. Last Accessed 15 Jan 2019
5. Bundesministerium für Verkehr, Bau und Stadtentwicklung: Allgemeines Rundschreiben Straßenbau Nr. 26/2010, Bonn (2010)
6. Forschungsgesellschaft für Straßen- und Verkehrswesen: Merkblatt für die Ausbildung und Zertifizierung von Sicherheitsauditoren für Straßen (MAZS), Köln (2009)
7. Forschungsgesellschaft für Straßen- und Verkehrswesen: Arbeitskreis 2.7.2 Merkblatt für die Ausbildung und Zertifizierung der Sicherheitsauditoren von Straßen (MAZS). https://www.fgsv.de/gremien/strassenentwurf/sicherheitsaudit/mazs/. Last Accessed 9 Jan 2019
8. Brannolte, U., Baselau, C., Fischer, L.: Ausbildung von Auditoren für das Sicherheitsaudit von Straßen im Außerortsbereich: - ein Erfahrungsbericht. Straße und Autobahn(08) (2004)
9. Professur Verkehrssystemplanung: Qualifizierung zu Auditoren für Außerortsstraßen und Ortsdurchfahrten: Ausbildungsplan (2019)

The Concept of the Software to Analyse Road Safety Statistics and Support Decision Making Process

Irina Makarova[1(✉)], Ksenia Shubenkova[1], Timur Bakibayev[2], and Anton Pashkevich[3]

[1] Kazan Federal University, Syuyumbike Prosp, 10a, Naberezhnye Chelny 423812, Russia
kamivm@mail.ru, ksenia.shubenkova@gmail.com
[2] Almaty Management University, Rozybakyiev Str., 227, Almaty 050060, Kazakhstan
[3] Politechnika Krakowska, Warszawska 24, 31-155 Kraków, Poland

Abstract. Specialists of different fields (from traffic management and control to logistics) are interested in the market of autonomous cars development. This could give a synergistic effect, because it can improve reliability, safety, efficiency and sustainability of the transport system. However, intellectualization of the transport system requires solving problems that arise (here, under the intellectualization we understand the process of transforming the transport system into the Intelligent Transport System). We propose the classification of risks that will arise while transport system's intellectualization. Classifier is made in accordance with risk characteristics that will allow directing efforts to prevent the most probable risks, as well as to reduce their severity in case of occurrence. We also suggest software based on a multiple-factor analysis of information that identifies the causes of critical situations. We have modified Haddon matrix, which allows determining factors that affect the number of accidents and the severity of their consequences, as well as the measures most effectively contributing to improving road safety. If recommendations' implementation caused increase of road safety, the proposed software enters these scripts into the knowledge base. This has allowed us to conclude what actions have helped to increase road safety in the city and what have to be corrected. If recommended decisions haven't had an expected positive effect, Haddon matrix was revised and adjusted in accordance with the actual results.

Keywords: Road safety · Transport system's sustainability · Risk analysis · Haddon matrix · Decision support system

1 Introduction

The environmental load created by the fast pace of urbanization both on the nature and on urban services leads to serious consequences [1]. However, density of cities can provide increased efficiency and technological innovation while reducing consumption of resources and energy. These problems' solution and the transition to sustainable and

A. Varhelyi et al. (Eds.): VISZERO 2018, LNITI, pp. 47–58, 2020.
https://doi.org/10.1007/978-3-030-22375-5_6

green economy is possible only through the transition to intelligent subsystems of the city, especially, in such important area as transport. Intelligent Transport System of the Smart City due to application of an integrated approach provides the opportunity to implement interactive services, which leads to a decrease in traffic congestion and makes the traffic convenient and safe for all its participants.

2 Road Safety in Smart City: Risks and Possible Solutions

Today most of accidents are connected with the so called "human factor". The research conducted by Zhang et al. [2] has proved that the category "unsafe behaviours" is the most frequent reason of traffic accidents. The category of "unsafe behaviours" includes such reasons of traffic accidents, as alcohol intake [3] and high speeding of vehicles [4]. M'bailara et al. [5] have also concluded that drivers' emotional status is a weighty factor influencing on the possibility of traffic accident's occurrence. Therefore, there is a widespread opinion that the semi-autonomous driverless vehicles will decrease the risk of accidents and will ensure sustainability and safety of the whole transport system. At the same time, according to Hicks [6], it should be taken into account that new technical and technological solutions can cause new problems which will require new methods and means to be mitigated and tackled. Hence, it is crucial to find the probable risks, determine the probability that they occur and what are their potential impacts. Moreover, the means of avoidance of these risks should and their consequences are supposed to be derived.

Transportation system risk analysis is such a profound problem owing to the unpredictable character of the possible accidents that may happen on the roads [7]. Optimization of the transport systems supposed to be analysed from the point of view of risk management. Risk management process consists of 4 parts [8]: (1) Risk identification – connected to recognition and classification of the probable risks that are associated with transportation systems management decision making process; (2) Risk assessment – association of probabilities of the risky situation occurrence in the system and the potential impact of those defined earlier; (3) Risk reduction – it is derived taking into consideration the probability and the weights of possible risk impacts. In order to control risks, the following methods can be applied: risk taking, risk mitigation, risk avoidance, risk transfer; (4) Risk monitoring – the measurement of the two phenomena: risk tracking and system changes fixing for punctual response and determination of the risk minimization measures effectiveness.

Different analytic methods devoted to the study of risks in transport system have been developed and modified in a number of studies: (a) way of rank determination for dangerous road locations by means of analytical hierarchy process (AHP) [9]; (b) modification of the EuroRAP model [10]; (c) method to determine the "bottlenecks" of the road network and the ways to measure their vulnerability [11]; (d) application of neural network and fuzzy logic for the means of increase the level of accident risk detection by using intelligent devices [12]; (e) the hierarchy of the levels of road safety risks is determined by applying data envelopment analysis (DEA) [13].

When identifying risks in transport system, it is necessary to take into account not only the possibility (probability) of technical failures occurrence, but also to assess the problems associated with transportation itself, as well as with the decision-making process. Risk assessment should be based on 2-D indicator set: risk probability and the level of damage consequences [14]. Since from the one hand, in case of transport system's risk management, such methods as risk avoidance and risk transfer are unacceptable, and on the other hand, it is rather difficult to develop actions to mitigate or prevent all possible risks due to limited resources, management actions have to be developed to prevent or mitigate the risks in accordance with their risk level obtained by multiplying risk probability to its consequences.

After identifying risks and their assessment, it is necessary to find solutions that reduce the risk of an accident and the severity of their consequences. One of the methods used to determine the factors associated with an accident is the model called the Haddon Matrix, which classifies the factors into three groups (human, automotive and environmental factors) across three time intervals – before the accident, during and after the accident [15]. Road transport has been called by Haddon a "man-machine" system. In addition, this system is poorly designed and needs to be cautiously treated. Crash factors can be analysed with assistance of Haddon matrix. Thanks to this analysis, it is possible to develop crash avoidance measures. The accident can be divided into 3 phases: pre-crash phase (crash prevention measures are undertaken then); crash phase (associatied with injury prevention measures) and post-crash phase (minimization of the crash impacts).

It is possible to prevent disasters and to reduce the severity of incidents only on the basis of multiple-factor analysis of information identifying the causes of critical situations. In our opinion, for these purposes it is possible to use such opportunities as OLAP and Big Data, which allows taking into account different groups of risks, for example, quality of algorithms, communications reliability by data transfer, cyber-threats. This is an effective tool that allows you to convert unstructured data into information that facilitates the adoption of management decisions. More detailed it was discussed in our previous paper [16]. Today, Big Data processing methods are constantly evolving due to the development of the Smart City concept. In order to collect, store and process information, decision support systems (DSS) are created, which are based on the intelligent analysis of data with the use of a wide range of modern scientific methods [17, 18].

3 The Proposed Decision Support System

DSS structure is usually dependent on the many different factors, for instance: kind of objectives to be tackled, available data, information and knowledge, etc. The proposed DSS contains three major parts: (1) A data system for gathering and preserving rule-form information obtained from internal and external sources; (2) A dialogue system allowing to set the data to be chosen and methods for their analysis; (3) A system of models (consisting of: ideas, algorithms and procedures) for data processing and analysis. Data processing can be performed by means of various approaches, from mere search to statistical analysis and nonlinear optimization. Since we are building a software mostly for data analysis, we decided to use Python on back-end. Server-side software stores data on PostgreSQL and serves it in JSON format. In the end, we get a

RESTful service that is capable of generating charts and doing any kind of analysis. Each record represents a traffic accident. Note that "description" field may include any kind of auxiliary information that is needed for analysis, such as "smartphone" or "bad weather". We suppose to keep raw natural text information in this field to not lose any information that we may need in the future. The developed program code of our software can be accessed in the Internet [19]. Since the back-end serves a RESTful service, the methods to CRUD data are presented in the Table 1.

Table 1. The used methods and codes.

Retrieve data:	
Method	GET
URL	/accidents/
Search	/accidents/?search = smartphone
Filter by date	/accidents/?date_from = 2018-01-01&date_to = 2018-12-31
Filter by time	/accidents/?time_from = 06:00&date_to = 14:00
Example	curl -X GET -H "Content-Type: Application/JSON" "/accidents/"
Add data:	
Method	POST
URL	/accidents/
JSON Body	{"date":__, "time":__, "description":__, "injured":__, "killed":__, num_vehicles:__, "num_persons":__, "malfunction":__, "alcohol": __, "pedestrians":__, "violation_code":__, "w_seat_belts":__, "w_o_seat_belts": __}
Example	curl -X POST -H –body '{"date": "2018-01-01", "time":"08:00"}' "Content-Type: Application/JSON" "/accidents/"
Change data of an accident with id = n:	
Method	PUT
URL	/accidents/n/
JSON Body	{"date":__, "time":__, "description":__, "injured":__, "killed":__, num_vehicles:__, "num_persons":__, "malfunction":__, "alcohol": __, "pedestrians":__, "violation_code":__, "w_seat_belts":__, "w_o_seat_belts": __}
Example	curl -X PUT -H –body '{"injured": 2}' "Content-Type: Application/JSON" "/accidents/n/"

The fields in our database are as follows (all these data are collected and stored for each section of the road network, and the user can get the information about every road section by requests):

```
region = models.IntegerField(default=0)
date = models.DateField()
time = models.TimeField()
description = models.TextField(max_length=5000)
injured = models.IntegerField(default=0)
killed = models.IntegerField(default=0)
num_vehicles = models.IntegerField(default=2)
num_persons = models.IntegerField(default=2)
malfunction = models.BooleanField(default=False)
alcohol = models.BooleanField(default=False)
pedestrians = models.IntegerField(default=0)
violation_code = models.CharField(max_length=10,
blank=True, null=True)
w_seat_belts = models.IntegerField(default=2)
w_o_seat_belts = models.IntegerField(default=0)
```

Malfunction and alcohol are Boolean fields that indicate that vehicle malfunction or alcohol caused the traffic accident. Most of the fields that are not present in our database are sensitive information that cannot be passed to our universities even under signed NDA. To work with data, we have created the special interface. It provides information input and its administration, though we import data from local police departments. Besides, users can create queries and obtain the necessary data for the data mining. The user interface is an HTML + Javascript code that consumes the REST service (Fig. 1).

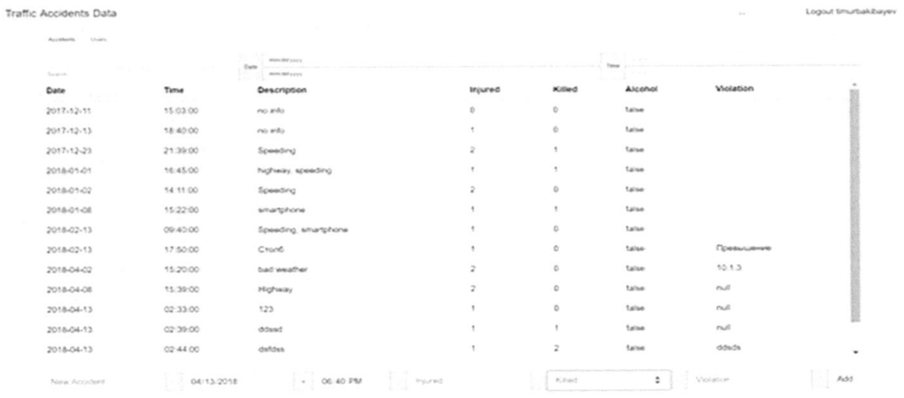

Fig. 1. User interface of the proposed software.

4 Results and Discussions

We propose the classification of risks accompanying intellectualization of the entire transport system and its subsystems (vehicles, infrastructure, means of control and informing, etc.). Classifier is made in accordance with risk characteristics: technical, environmental, organizational, economic, legal and ethical and social (Table 2). This will allow directing efforts to prevent the most probable risks, which can have the most negative consequences, and also reduce their severity in case of occurrence.

Table 2. Potential risks accompanying transport system's intellectualization.

Index	Risk	Risk probability	Conse-quences	Risk level	Ways of influences
Technical					
1.	Reduction of operational reliability because of increased complexity of the vehicles' design	3	3	**9**	Service system improvement, the use of highly reliable components, redundancy of elements that ensure safety
2.	Increased infrastructure requirements	4	4	**16**	The use of new technologies and high-strength materials when constructing
3.	Increased requirements to the weather conditions	3	4	**12**	Increasing the sensitivity of the sensors
4.	Increased requirements to communication systems	4	5	**20**	Development of new communication systems, duplication of communication channels
Ecological					
5.	The risk of technological disasters if there are cyber attacks or failures in the control system	2	5	**10**	Development of information security systems, improving the reliability of control systems, implementation of backup systems
6.	Increased negative impact on environment because of expansion of the vehicles' fleet	3	3	**9**	The use of resource-saving technologies, improving the efficiency of

(*continued*)

Table 2. (*continued*)

					transportation management
7.	The risk of road networks' overload due to the redistribution of traffic flows by modes of transport	3	3	**9**	Increasing the attractiveness of "green" modes of transport, improving the quality of transportation planning
Organizational					
8.	Complexity of the movement algorithms for rough terrain	2	2	**4**	Improving of the movement algorithms and the vehicles' design
9.	Disorientation in bad weather conditions	3	3	**9**	Improving of position control systems
10.	The risk of software hacking	3	4	**12**	Development of effective protection systems
11.	The absence of a panoramic view of the streets, impeding the routing	3	3	**9**	Financing development of effective GIS systems and updating maps
12.	Increased requirements to information processing speed	4	4	**16**	Support for the companies involved in development of technical and software solutions in the field of Big Data
13.	Complexity of decision-making in unusual situations	3	5	**15**	Development of expert systems and knowledge bases, accidents' statistics collection
14.	Complexity of communication with traditional nonautonomous vehicles	4	3	**12**	Improvement of gesture recognition and speech-understanding technologies
Economic					
15.	High cost of infrastructure changes	4	3	**12**	Development of high-strength materials and durable construction technologies
16.		5	2	**10**	

(*continued*)

Table 2. (*continued*)

	The high price of the vehicles				Support for corporate vehicle fleets to expand the market
17.	Increased total travel time due to the expansion of potential consumers	2	3	**6**	Improvement of transportation planning system, development of bonus and penalty schemes
18.	Redistribution of traffic and passengers flows by modes of transport	2	2	**4**	Regulation of transportation tariffs, promoting economic, eco-friendly and safe modes of transport
Legal and ethical					
19.	Loss of privacy	4	3	**12**	Improvement of personal data security system
20.	Adequacy of algorithms for critical situations	4	5	**20**	Analysis of accumulated statistical information on accidents and improvement of algorithms
21.	Ambiguity of legal responsibility for damage	4	3	**12**	Legislation improvement
22.	Ambiguity of legal responsibility when organizing transportation	4	4	**16**	Legislation improvement
Social					
23.	Loss of self-driving capability	2	2	**4**	Allocation of special roads for drivers of traditional cars
24.	Lack of driving experience of drivers in critical situations	4	5	**20**	Improvement of drivers and autonomous cars' users training, creation of malfunction protection system
25.	The loss of jobs by people whose work is related to driving vehicles	3	3	**9**	Retraining of drivers, provision of alternative jobs
26.	Possibility of mining the vehicle	3	5	**15**	Security upgrade, temperproof systems

Table 2 has been prepared taking into account opinions of ten experts (employed in the City Transport Division and private carriers, working in the transport sphere not less than 5 years), who were asked to assess the possibility of every risk according to the gradation: 1 – a very low probability; 2 – a low probability; 3 – risk is probable; 4 – a high probability; 5 – a very high probability. Potential damage risk evaluations have been performed according to the following consequences impact levels: 1 – without consequences; 2 – with small consequences; 3 – with great consequences; 4 – with critical consequences; 5 – with catastrophic consequences. Results have been multiplied and included in the "Risk level" column. These risks can be considered not only for the city under examination, hence, the presented approach and DSS can be spread to other cities. We plan to check our DSS in other cities with different spatial planning, different square and different motorization level.

Analysis of the risks' levels shows that the most probable risks with serious consequences for both human and the transport system in general are related to the quality of information and communication technologies (including software, communications systems, security systems, etc.). Therefore, we have modified the Haddon matrix (Fig. 2) by including the concept of "information environment" to more general concept "environment". This will allow identifying countermeasures that need to be taken to prevent threats both in the short and long term.

Having implemented the factors presented in Haddon matrix, the following stage is to assess their efficiency.

One of the examples can be usage of cameras detecting speed limits exceeding. It resulted in decrease in over-speeding. By using of such cameras, further measures can be also undertaken to reduce the possibility of an accident in a given area. The decision making process in our developed software is the following. The stored in the database data for each section of the road network are analyzed by different parameters in comparison with previous periods. The DSS takes into account all the infrastructure and organizational changes that were implemented during this period. If the number of accidents and violations in this road section has decreased, the DSS concludes that the managerial decisions taken are correct and enters these decisions into the knowledge base so that next time, on another road section with similar problems, to recommend a certain set of activities. If the safety indicators of this road section have decreased, the system concludes that the measures taken are incorrect and offers other solutions. Thus, our DSS is based on the feedback principle. This means that it analyzes the consequences of decisions taken and adjusts its recommendations in accordance with the accumulated experience. With the use of the proposed software, we have compared the accidents' statistics for last years (Fig. 3). Preliminary results have shown that there was a diminution in the amount of accidents happened in pedestrian crossings in 2016. Other reason-based accidents have also been decreased (except for the ones associated with leaving the oncoming traffic). Moreover, it is also a result of sanctions becoming more and more strict. Furthermore, in 2016 there were speed bumps installed in the areas of unregulated pedestrian crossings. As a result, the traffic flow speed has been reduced there what lead to a smaller accident probability.

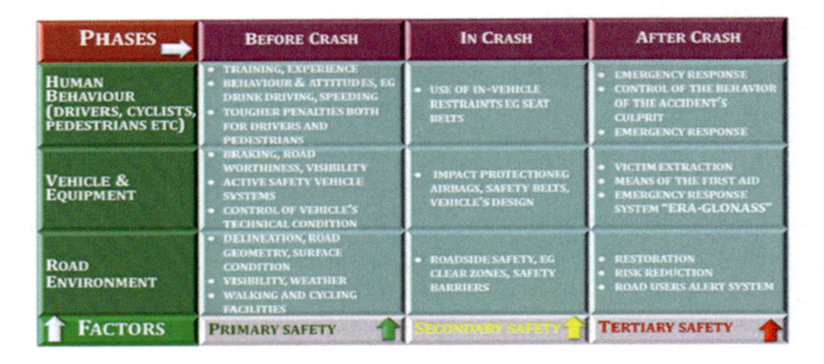

Fig. 2. Modified Haddon matrix.

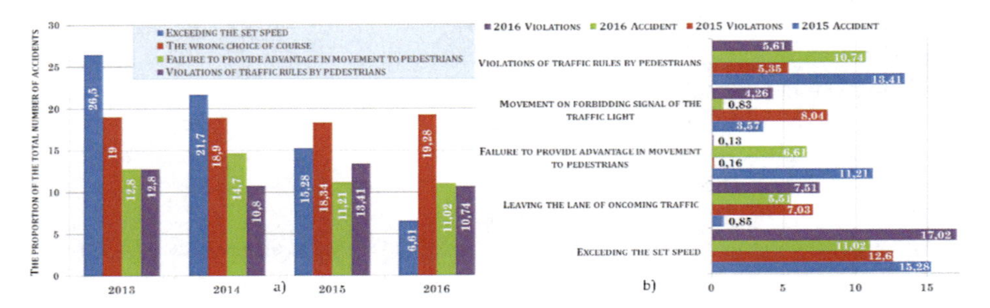

Fig. 3. (a) The causes of the road accidents; (b) The relationship between traffic violations and number of road accidents.

5 Conclusion

The main idea of the software proposed is that all managerial decisions made in such complex systems as transport one, have to be analysed from the point of view of their efficiency. For this purpose, the DSS should compare statistics of road accidents not only before the decision making process, but also after the suggested decisions are implemented. At the current stage of our research we have created the special software for information input and analysis. One of the most dangerous pedestrian road section (according to the 2015 statistics of the road accidents with pedestrians injures) was taken into consideration as an object under study. The most common cause of the road accidents on this road section was the exceeding of the speed limits. On the base of these results, municipal State Traffic Inspectorate made a decision to install the speed bumps before the pedestrian crossing. After the year, the statistics of the road accidents with pedestrians injures at the same road section was analysed once again. The reduced number of accidents caused by exceeding the speed limits has proved the decision made and it was entered into the DSS knowledge base as one of the possible solutions.

The next step of our research is to form the knowledge database on the basis of results of decisions taken. Since the analyses have to be complex, taking into account also infrastructural peculiarities, as well as the intensity and structure of the traffic flow of road sections, in our future work we will continue developing our DSS, extending it by simulation models and implementing Haddon Matrix modified by including the concept of “information environment”.

Acknowledgements. Research is partially funded by national grant No. BR05236644.

References

1. Pernebekov, S.S., et al.: Modeling of traffic flows with due regard to ecological criteria. Life Sci. J. **11**(5), 300–302 (2014)
2. Zhang, Y., et al.: New systems-based method to conduct analysis of road traffic accidents. Transp. Res. Part F: Traffic Psychol. Behav. **54**, 96–109 (2018)
3. Hedlund, J., et al.: Pre-offense alcohol intake in homicide offenders and victims: a forensic-toxicological case-control study. J. Forensic Leg. Med. **56**, 55–58 (2018)
4. Ramana, G.V., Reddy, N.S., Praneeth, D.M.V.: Analysis on traffic safety and road accidents: a case study on selected roads in Muscat, Oman. Int. J. Civ. Eng. Technol. **9**(2), 417–427 (2018)
5. M’bailara, K., et al.: Emotional reactivity: beware its involvement in traffic accidents. Psychiatry Res. **262**, 290–294 (2018)
6. Hicks, D.J., et al.: The safety of autonomous vehicles: lessons from philosophy of science. IEEE Technol. Soc. Mag. **37**(1), 62–69 (2018)
7. Xie, G., et al.: Situational assessments based on uncertainty-risk awareness in complex traffic scenarios. Sustain. (Swit.) **9**(9), 1582 (2017)
8. IEC/ISO 31010: Risk management – risk assessment techniques (2009). https://www.iso.org/standard/51073.html. Accessed 14 Jan 2019
9. Agarwal, P.K.: A methodology for ranking road safety hazardous locations using analytical hierarchy process. Proc. – Soc. Behav. Sci. **104**, 1030–1037 (2013)
10. Development of Risk Models for the Road Assessment Programme, 46: http://www.eurorap.org/wp-content/uploads/2015/04/20120308-website-sdl-DAL-TRL_RAP_Model_Development_Final_3.pdf. Accessed 14 Jan 2019
11. Vulnerability and Risk Analysis of Road Infrastructure in Reykjavik, 41: http://www.vegagerdin.is/vefur2.nsf/Files/Vulnerability_riskanalysis-PrelimReport/$file/Vulnerability_risk%20analysis-PrelimReport.pdf. Accessed 14 Jan 2019
12. Beinarovica, A., Gorobetz, M., Levchenkov, A.: Innovative neuro-fuzzy system of smart transport infrastructure for road traffic safety. IOP Conf. Ser.: Mater. Sci. Eng. **236**(1), 012095 (2017)
13. Shah, S.A.R., et al.: Road safety risk assessment: an analysis of transport policy and management for low- middle- and high-income Asian countries. Sustain. (Swit.) **10**(2), 389 (2018)
14. Makarova, I., Shubenkova, K., Gabsalikhova, L.: Analysis of the city transport system’s development strategy design principles with account of risks and specific features of spatial development. Transp. Probl. **12**(1), 125–138 (2017)
15. Haddon Jr., W.: Advances in the epidemiology of injuries as a basis for public policy. Public Health Rep. **95**, 411–421 (1980)
16. Makarova, et al.: Application of data mining technology to optimize the city transport network. IEEE AICT 2016, Baku, Azerbaijan, paper #7991674 (2016)
17. Behnood, H.R., et al.: A fuzzy decision-support system in road safety planning. Proc. Inst. Civ. Eng.: Transp. **170**(5), 305–317 (2017)

18. Ryder, B., et al.: Preventing traffic accidents with in-vehicle decision support systems - the impact of accident hotspot warnings on driver behavior. DSS **99**, 64–74 (2017)
19. Software Source Code: https://github.com/timurbakibayev/ta. Accessed 14 Jan 2019

Analysis of Road Traffic Safety Increase Using Intelligent Transport Systems in Lithuania

Aldona Jarašūnienė(✉) and Nijolė Batarlienė

Vilnius Gediminas Technical University,
Plytinės str. 27, 10105 Vilnius, Lithuania
{aldona.jarasuniene,nijole.batarliene}@vgtu.lt

Abstract. Vehicle congestion and high risk of accidents result in disappointment of many drivers and public transport users, causing psychological tension and increasing the accident rate. The accident rate is one of key indicators that help measuring processes of the transport system and its development. With the help of various road safety solutions, the modern world aims to ensure safe transportation and achieve reduction in the number of traffic accidents. Lithuania is focusing on these issues as well. Reduction in the number of traffic accidents requires implementation of the national road traffic safety policy. Traffic safety on the roads of national significance in Lithuania is one of the most important priorities. The success of reducing the number of victims in road accidents lies in mutual understanding, respect, cultured behaviour, closer cooperation between traffic members, as well as installation of engineering safety measures and application of advanced technologies. In order to properly evaluate the impact of Intelligent Transport Systems (ITS) on traffic safety, the article presents a comprehensive statistical analysis on traffic accidents, ITS applications on Lithuanian roads, measures to minimize accident levels by deploying Intelligent transport systems.

Keywords: Traffic · Safety · ITS · Accident · Information

1 Introduction

The Intelligent Transport System (ITS) works with information and control technologies providing the core of ITS functions. ITS services can make transport safer and more secure. The purpose of ITS is to collect information about traffic flows and conditions on roads and to present the obtained data for control systems.

As global practices show, intelligent transport systems may have wider areas of application. The highest efficiency of ITS would be achieved if a unified system connecting all modes of transport, multimodal transport infrastructure and traffic control is established. Data and information should be transferred from one vehicle to the other (V2X), as well as information is to be shared between vehicles and infrastructure and vice versa (V2I), and from one infrastructure to the other (I2X).

In 2011, March 2, a new national road safety programme was approved to implement the vision of safe traffic and so that no road users are killed or suffer serious injuries in Lithuania. The strategic objective of the programme is to attain that

A. Varhelyi et al. (Eds.): VISZERO 2018, LNITI, pp. 59–66, 2020.
https://doi.org/10.1007/978-3-030-22375-5_7

Lithuania appears between the 10 European Union Member States with the best results in terms of the road users killed in traffic accidents per 1 million of the country's inhabitants (or no more than 60 people killed in traffic accidents per one million of inhabitants), by improving the state of road safety. Several priorities are set to attain this objective and one of it is the modern information technologies. The objective of the priorities is to improve the process of collecting and presenting traffic data and implementing and developing Intelligent Transport Systems (ITS).

The estimation on traffic safety problems in roads of Lithuania using intelligent transport system is examined in the article. The main problems are identified, and they are solved by the ways with ITS.

2 Literature Review

Although ITS activities have been carried-out in Europe for a long-time, the majority of these systems are innovative solutions and the information gathered on it is not sufficient to assess impact; development activities are uncoordinated and fragmented. The majority of ITS do not have the assessment of the impact on traffic safety due to short-term exploitation, and other factors simultaneously affecting traffic safety which impede the objective evaluation of the system. Objective comparisons between ITS are difficult due to these reasons.

ITS activities are based on systems, the basis of which are comprised of a well-developed monitoring equipment network of weather and traffic conditions, information centres and various driver information [8].

Intelligent transport systems are the aggregate of information and telecommunication devices ensuring a safe and effective movement of vehicles, cargos and people in road sector [14]. ITS help to manage transport flows by reducing congestions, downtime as well as increase safety by warning about a possible accident situation due to the personal or environmental impact [4]. One of the measures, increasing the impact of road traffic safety is ITS development. ITS include many activities, among which are the monitoring of road traffic conditions and informing of traffic participants about those conditions [2].

Using of Intelligent Transport Systems have been analysed by Bekiaris and Nakanishi [3], Chowdhury and Sadek [4], Ezzel [6], Jarašūnienė [8], Sussman [20], Hancock and Xu [13], Li and Miao [18], Schneider et al. [19], Kabashkin [11].

3 Investigation of Traffic Safety Situation on the Roads of Lithuania

3.1 Traffic Safety Situation on the Roads of the State of the Republic of Lithuania

Over the past 25 years, the number of vehicles on Lithuanian roads doubled. The road network was not adapted to such flows, which led to a deterioration of the road safety situation. The improvement of accident rates has not been achieved for a long time, however, upon joining and agreeing to certain obligations to the EU, the number of fatalities in Lithuania decreased more than double. Such a result is determined by the intensive applications of engineering measures to attain traffic safety as well as social initiatives organized by the Lithuanian Road Administration under the Ministry of Transport and Communications. However, even such an improvement of road safety situation is not satisfactory as the ultimate goal is a zero vision in which no one is killed or badly injured on roads. In order to clarify the road safety situation, it is important to provide data on the number of vehicles and fatalities, road accidents and the ones injured, as well as the most dangerous roads of the country, black spots, etc.

During the period of application of the national road safety program, the number of fatalities per 1 million of the country's population decreased by 27%. In 2017, there were 3,192 traffic accidents on Lithuanian roads and streets, with 192 fatalities and 3,752 injuries. In comparison to 2016, the numbers of accidents and victims have barely changed: accidents decreased by 0.3%, injuries increased by 0.1%, and the number of fatalities remained the same.

Upon the new registration procedures introduced at the beginning of July 2014, vehicles that failed to comply with the compulsory civil insurance and/or technical inspection requirements were unregistered and the vehicle fleet decreased by more than a third compared to the previous year. In 2017 Lithuanian vehicle fleet increased by 3.75% – 60,613 cars in comparison with 2016. Accordingly, the number of vehicles increased from 566 to 596 per 1,000 inhabitants. Figure 1 shows a statistic of vehicles and fatalities 1998–2017.

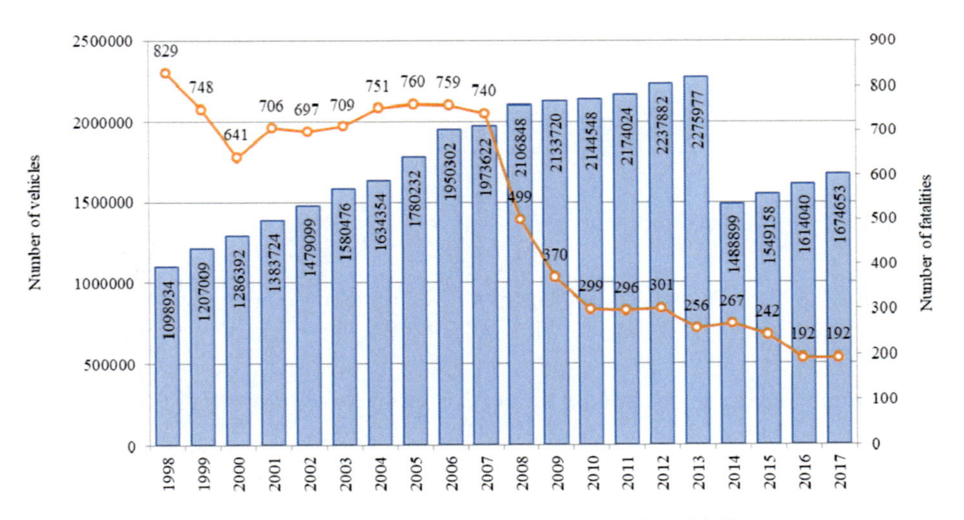

Fig. 1. Number of vehicles and fatalities 1998–2017.

In 2017 Lithuanian vehicle fleet increased by 3.75% in comparison with 2016, while the number of fatalities in Lithuanian roads did not change.

3.2 Traffic Safety Situation on the National Roads of Lithuania

In 2017, 444 road accidents occurred on the roads of Lithuania, with 59 road users killed and 607 injured. The following Table 1 gives numbers of traffic accidents on national roads 2014–2017.

Table 1. Traffic accidents and victims on national roads 2014–2017.

Year	Traffic accidents	Killed	Injured
2014	419	60	539
2015	449	74	592
2016	419	44	561
2017	444	59	607

Figure 2 graphically shows traffic situation on the national roads of Lithuania.

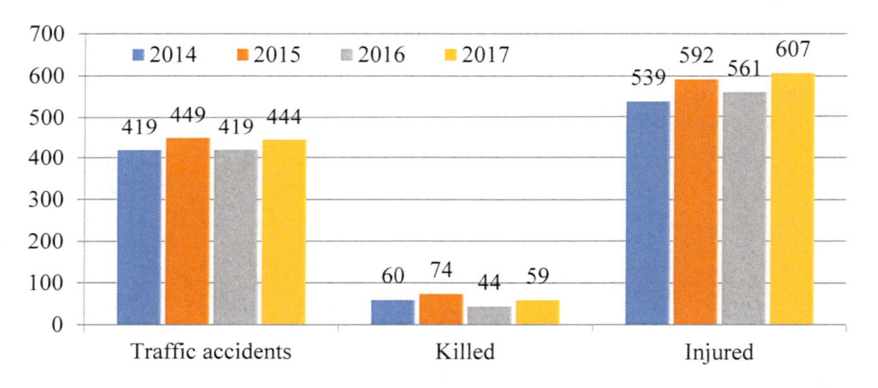

Fig. 2. Traffic accidents and victims on national roads 2014–2017.

In 2017, traffic accidents accounted for 44% of accidents emerging on the roads of national significance and 14% – for all road accidents in the country.

3.3 Traffic Accidents and Victims on the European Road Network (TEN-T)

The Trans-European Transport Network (TEN-T) consists of the comprehensive TEN-T network; the core TEN-T network; and corridors of the major network. The length of the core TEN-T network in Lithuanian territory is 623,476 km. In 2017, the comprehensive TEN-T network had 16 traffic accidents in which 3 people died and 20 were injured, whilst the core TEN-T network had 18 traffic accidents in which 4 people died and 20 were injured.

In 2017, there were 22 black spots identified on the roads of national significance in Lithuania. 8 black spots were detected on the main roads, 11 – on the national roads and 3 black spots were detected on the country lanes. In comparison with 2016, there were 9 new black spots determined on the roads of national significance in 2017, 11 black spots disappeared, and 13 black spots remained. There were 111 traffic accidents in which 16 people died and 135 were injured on the identified black spots of the roads of national significance during the period 2014–2017.

Black spot – a road segment of 500 m in which no less than 4 traffic accidents occurred during the four-year period and accident indicators (density of accidents and accident rate1) have reached or exceeded the limit values. These places are usually at one-level intersections with limited visibility, bus stops, low visibility road sections with horizontal and vertical curves, road sections with trees growing close to the lanes, and other road sections in which some of the road elements poses a higher risk of traffic accident.

Figure 3 shows a situation of black spots on the regional roads of Lithuania, and Fig. 4 shows a situation of black spots on highways.

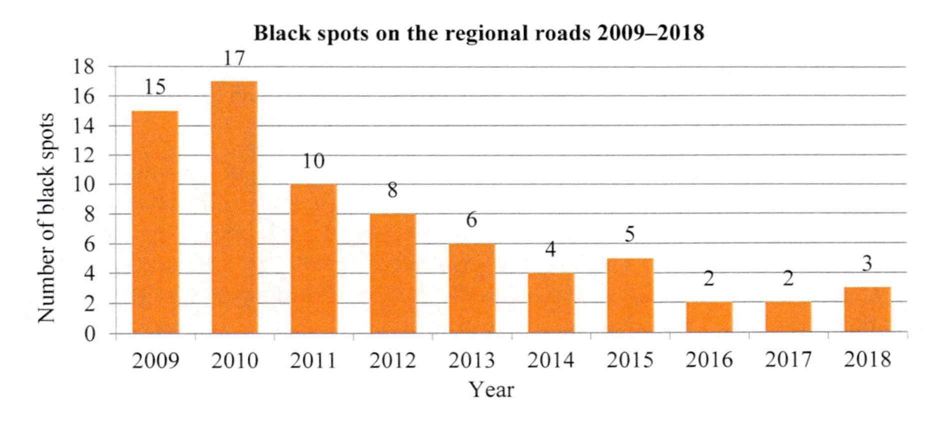

Fig. 3. Fluctuation in the number of black spots on the regional roads 2009–2018.

Accident density indicates the number of road accidents per 1 km of road section per year. Accident rate indicates the number of accidents per 1 million cars passing 1 km of road a year [16].

4 Measures to Reduce Accidents Rates by Using Intelligent Transport Systems

Automatic Vehicle Braking System. A great amount of accidents occurs due to delayed driver reaction or insufficient braking power. This can be explained by distraction, inattention, poor visibility or unexpected obstacle on the road. Several manufacturers have developed the Autonomous Emergency Braking (AEB), which can help the driver to avoid these types of accidents or at least reduce its consequences. These autonomous systems operate independently of the driver. This technology is

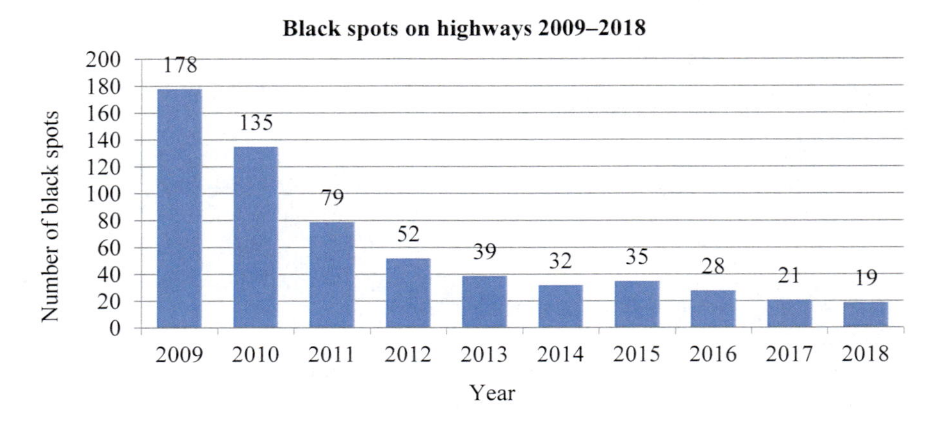

Fig. 4. Fluctuation in the number of black spots on highways 2009–2018.

based on the radar, camera and laser sensors to detect the risk of collision. It can even detect if the car in the front is slipping or braking. If the threat of an accident occurs, the braking system is activated automatically.

Lane Support System. Various systems for autonomous vehicles are being developed to minimize human misconduct while driving. One of these systems is the lane support system. The system is activated only at higher speeds on the countryside and only when no turn signal is displayed. The lane support system operates in the same way as the lane warning system, only in this case not only the lane departure alert is activated, but also corrective actions are automatically performed to get vehicle back into the lane. If the vehicle approaches the horizontal marking line very close, the system gently turns the vehicle towards the middle of the lane by gently braking any of the wheels or directly turning the steering wheel.

Blind Zone Monitoring System. All vehicle manufacturers are faced with the problem of improving the visibility of the blind zone. Currently developed blind zone monitoring systems can be classified into active and passive. The active monitoring system uses diverse electronic surveillance devices mounted on the exterior of the rear-view mirrors or near the rear bumper of the vehicle. This surveillance equipment may capture vehicles passing by the help of the electromagnetic waves (radar operation principle) or digital cameras for real-time monitoring and analysis. If the driver turns signal and one of these devices detects another vehicle within the blind zone, a visual and audible warning about the potential danger is sent to the driver. In dangerous situations more advanced systems do not allow to turn the steering wheel. Blind Zone Monitoring System has been continuously improved and tested.

Driver Attention Tracking System. Research has shown that about 20% of accidents occur due to driver fatigue. Driving for long distances and on straight lines adversely affect driver's psychological state. A monotonous image and a state of rest while sitting behind the wheel encourages a micro sleep. To solve this problem, certain technologies to detect driver drowsiness have been developed. The Driver Attention Tracking

Systems use a variety of technologies: digital camera monitors driver's face, temperature sensor captures the temperature of human face, and analyses change; other technologies include eye pupil and flashing type observation with special goggles, or steering motion analysis. As soon as the device detects drowsiness of the driver, the system alerts and warns it by sound signals or vibrations. To achieve greater efficiency, this system can be equipped with the Lane Support System. Driver Attention Tracking System has been continuously improved.
Since joining the EU and agreeing to certain obligations, the number of fatalities in Lithuania decreased more than double. Such a result is determined by the intensive applications of engineering measures to attain traffic safety as well as social initiatives organized by the Lithuanian Road Administration under the Ministry of Transport and Communications.

The highest efficiency of ITS would be achieved if a unified system connecting all modes of transport, multimodal transport infrastructure and traffic control is established. Data and information should be transferred from one vehicle to the other (V2X), as well as information is to be shared between vehicles and infrastructure and vice versa (V2I), and from one infrastructure to the other (I2X). The deployment of the ITS in transport infrastructure would result in multiple benefits for traffic safety, road quality, more efficient logistics and public transport services.

5 Conclusions

1. Since joining the EU and agreeing to certain obligations, the number of fatalities in Lithuania decreased more than double. Such a result is determined by the intensive applications of engineering measures to attain traffic safety as well as social initiatives.
2. In 2017, traffic accidents accounted for 44% of accidents emerging on the roads of national significance and 14% – for all road accidents in the country.
3. In 2017, there were 22 black spots identified on the roads of national significance in Lithuania. 8 black spots were detected on the main roads, 11 – on the national roads and 3 black spots were detected on the country lanes. In comparison with 2016, there were 9 new black spots determined on the roads of national significance in 2017, 11 black spots disappeared, and 13 black spots remained.
4. Although ITS activities have been carried-out in Europe since the 1980's, the majority of these systems are innovative solutions and the information gathered on it is not sufficient to assess impact.
5. Aiming at the tackling of traffic safety problem, it is necessary to implement national traffic safety improvement politics. Traffic safety insurance should be the main priority.
6. It has been proved that traffic safety improvement requires perception and evaluation of the main determinants and possibility to influence them.

References

1. Barfield, W., Dingus, T.A.: Human Factors in Intelligent Transportation Systems. Lawrence Erlbaum, Mahwah (1999)
2. Batarlienė, N.: Informacinės transporto sistemos, 335 p. Technika, Vilnius (2011)
3. Bekiaris, E., Nakanishi, Y.J.: Economic Impacts of Intelligent Transportation Systems: Innovations and Case Studies. Elsevier, Amsterdam (2004)
4. Chowdhury, M.A., Sadek, A.W.: Fundamentals of Intelligent Transportation Systems Planning. Artech House, Norwood (2003)
5. Elvik, R., Hoye, A., Vaa, T., Sorensen, M.: The Handbook of Road Safety Measures, 2nd edn. Emerald Group Publishing, Bingley (2009)
6. Ezzel, S.: Intelligent Transportation Systems. The Information Technology and Innovation Foundation, Washington (2010)
7. Harvey, J., Shaw, M., Shaw, S.L.: GIS-T DATA MODELS
8. Jarašiūnienė, A.: Intelektualiosios transporto sistemos. Technika, Vilnius (2008)
9. Jarasuniene, A.: Information system using in transport institution. In: Proceedings of 9th International Conference "Transport Means 2005", pp. 133–136 (2005)
10. Jarašūnienė, A., Miliauskaitė, L.: The estimation on traffic safety problems in road Lithuania using ITS. In: RelStat 2011: Proceedings of the 11-International Conference in Riga, pp. 327–334 (2011)
11. Kabashkin, I.V.: Main Activities in Latvian Transport Sector of the Intelligent Transport Systems Programs. RAU Scientific Reports. Computer Modeling & New Technologies 2: 61–65 p. Riga Aviation University, Lomonosov Str.1, Riga, LV-1019, Latvia (1998)
12. Kachroo, P., Özbay, K.: Feedback Ramp Metering in Intelligent Transportation Systems. Springer, New York (2003)
13. Hancock, K., Xu, J.: Modeling Regional Freight Flow Assignment Through Intermodal Terminals, 26 p. (2005)
14. Kirikova, M., Grundpenkis, J., Wojtkowski, W.: Information Systems Development: Advances in Methodologies, Components, and Management. Springer, New York (2002)
15. Marma, A., Eidukas, D., Žilys, M., Valinevičius, A.: Intelektualiųjų transporto valdymo sistemų efektyvumas. ISSN 1392 – 1215 Elektronika ir elektrotechnika **6**(62): 61–66 (2005)
16. Methodology for Identification of Dangerous Segments on the Roads of National Significance, approved in 2011. June 7 Order of the Minister of Transport and Communications of the Republic of Lithuania No. 3–342
17. Anderson, M.D., Souleyrette, R.R.: Simulating traffic for incident management and ITS investment decisions. In: Transportation Conference Proceedings, pp. 7–10 (1998)
18. Li, Q., Miao, L.: Integration of China's intermodal freight transportation and ITS technologies. In: Proceedings of the 2003 IEEE International Conference on Intelligent Transportation Systems, vol. 1, pp. 715–719. IEEE (2003)
19. Schneider, W., Asamer, J., Mrakotsky, E., Toplak, W.: Influence of environment conditions on traffic flow. In: Proceedings of the 2007 IEEE Intelligent Transportation Systems Conference, Seattle, WA, USA, September 30–October 3 (2007)
20. Sussman, J.M.: Perspectives on Intelligent Transportation Systems (ITS). Springer, New York (2005)
21. Transportation Research Board. Standards for Intelligent Transportation Systems: Review of the Federal Program (2000)
22. World Road Association. Road safety manual

Dangerous Goods Transport Problems in Lithuania

Nijolė Batarlienė(✉)

Vilnius Gediminas Technical University,
Plytinės str. 27, 10105 Vilnius, Lithuania
nijole.batarliene@vgtu.lt

Abstract. Dangerous goods are a specific part of all goods. About 50% of cargoes are considered dangerous when they are transported in the Republic of Lithuania. Anyone transporting or storing dangerous goods must solve two additional problems: choose the mode of transport and reduce the risk of accidents and the potential damage to people and the environment during transport.

Carrying goods by road involves the risk of traffic accidents Dangerous goods that are transported or stored incorrectly can cause human or animal disease, poisoning, burns, as well as explosion, fire, damage to other cargo, rolling stock, structures and equipment, contamination of the environment and water.

This article analyzes accidents and incidents of dangerous goods transportation by road transport. There are presented the research results indicating the way the respondents estimate the main factors related to risk in road transport, there is provided analysis thereof. The main causes for traffic accidents incurred by enterprises and factors determining route selection are presented. Recommendation measures for reduction of accidents in road transport while transporting dangerous goods are presented.

Keywords: Dangerous goods · Risk · Control · New technologies · Vehicle

1 Introduction

Dangerous goods have known as hazardous materials include flammable, explosive, corrosive, oxidizing, toxic, radioactive, or pathogenic. Dangerous goods can cause accidents and lead to fires, explosions and chemical poisoning or burning with considerable harm to people or the environment.

50% of all cargoes transported in Lithuania are dangerous goods. 70% of these cargoes are in transit. From them: 54% of dangerous goods are transported by rail, 25% of dangerous goods transported by road, 20% of dangerous goods transported by water, 1% of dangerous goods transported by air.

Accidents during the transportation of dangerous goods often have serious consequences: the socio-economic cost of an accident with dangerous goods may be twice as high as that of a "normal" goods-transport accident due to the dangerous goods escaping and the environmental damage caused by this. However, compared with the accident occurrence in the transportation of general goods, accidents involving dangerous goods

A. Varhelyi et al. (Eds.): VISZERO 2018, LNITI, pp. 67–73, 2020.
https://doi.org/10.1007/978-3-030-22375-5_8

are rare: around eight out of 1,000 personal injury accidents involving a goods vehicle are classified as accidents involving dangerous goods [9].

One problematic aspect of these accidents is that the cargo is often inadequately secured. This can be seen through analysis of "serious" accidents involving dangerous goods vehicles in which people are injured due to the dangerous goods or in which more than 100 kg or litres of the dangerous goods are released. As approximately 29% of dangerous goods transported are "poisonous", it is particularly important that the cargo is correctly secured, and the vehicles are correctly labelled.

Transportation of dangerous cargo is a complex and demanding process. The success of cargo transportation depends on the total number of personnel, in particular, the ones that are involved in loading, unloading processes, as well as ensuring that cargo is transported in accordance to all safety requirements [2].

When dangerous goods are carried, there is a risk of an incident or accident. An explosion, fire, chemical burn or environmental damage may occur. Most goods are not considered sufficiently hazardous to require special precautions during carriage. Some goods, however, have properties which mean they are potentially hazardous if carried.

In various articles, the authors present international co-operation opportunities for the prevention of accidents of dangerous goods [4, 5], present system of road transport agents for dangerous goods in risk management [6]. Other authors seek to tackle many problems of integrating real-time data information about the tracking of a dangerous material vehicle with risk evaluation methodologies in order to describe possible accident scenarios [8].

The goal of the article is to analyse incidents and accidents in the transportation of dangerous goods by road, to identify key risks related to transportation of goods as well as to assess safety aiming at reduction of incident risk.

2 Analysis of Research Results

The risks assessment is particularly important in the area of transport. This assessment can be carried-out by considering certain risk factors and enables to determine the route, calculate the possible costs of the consequences and other factors. For instance, risks can be minimized by manufacturing more firm packaging, doing special preparations for vehicle itself, educating drivers on safe transportation of dangerous goods.

Risk assessment allows to substantiate and prove the connection between individual technical elements and the impact on the overall technological process of transportation. Additionally, risk assessment enables to strengthen and improve the security of transport through existing technical measures and helps to search for newer and more effective measures.

To find out which problems are encountered with and which preventive measures are taken to reduce risks associated with transportation of dangerous cargo by road in Lithuanian transport companies, 127 transport companies that are liable to the ADR treaty were interviewed. 61% of the respondents filled in the questionnaire.

According to answers obtained from 78 respondents, the ADR cargo organization is the main activity in 61 companies while 5 companies do not perform such transportation and 14 companies claim that this type of activity is secondary in their company.

Therefore, the greatest attention is to be paid to 61 companies involved in dangerous cargo transportation.

A questionnaire conducted has shown that the majority of organizations transport Class 2 Gas, Class 3 Flammable Liquids, Class 8 Corrosive Substances, and Class 9 Miscellaneous Dangerous Goods.

The questionnaire covers questions on professional competencies of a person responsible for DG (dangerous goods) transportation in an organization. 61% of respondents claimed that there is a security specialist who holds a certificate proving his/her qualifications. 6% of organizations which are currently not engaged in the transport of dangerous goods, however has employed a specialist with a professional competence in the transport of dangerous goods. It can be assumed that companies have not refused to organize the transport of dangerous goods but are anticipating profitable orders and annual contracts.

Companies engaged in organizing dangerous cargo transportation must to reduce risks as much as possible. Due to inherent and uncontrolled factors, the elimination is not possible, however, companies must aim for risks value to be close to zero. Therefore, an open question "What measures does company use to reduce risks?" was included in the questionnaire. 7 companies claimed that they accept dangerous cargo to be transported only from reliable companies. Respondents from 8 companies noted that they reduce risks by sending their drivers to special trainings, 9 organizations use the ITS and observe vehicle movement in real-time, whereas 5 companies do not reduce risks as they do not consider that these preventive measures may impact their activities. 13 companies have indicated that they are trying to comply with all requirements aimed at transporting dangerous goods, 15 organizations are hiring only highly skilled employees. The remaining companies invest in purchasing new vehicles. The Fig. 1 shows a detailed distribution on the measures by which companies try to reduce their risk.

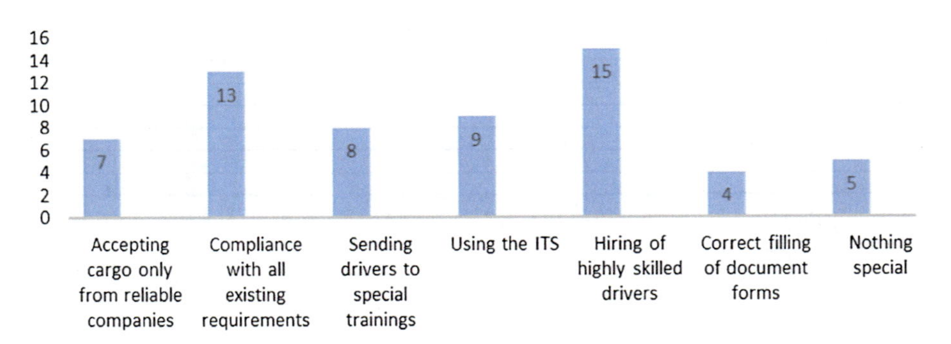

Fig. 1. Risks reduction measures of dangerous cargo transportation.

Scientific articles indicate that carriers usually reduce the risk of transportation by exploiting vehicles in a good technical condition, which means that technical preparation of vehicles transporting dangerous goods by road is not properly managed. According to the Causal Hierarchy, Lithuanian carriers identify the driver and his/her activities as the second most important factor determining the number of accidents on the roads, whilst the road and the environment is the third most important factor. As a result of this distribution, it becomes clear that, all of these factors are important to prevent accidents, however, technical vehicle condition receives a greater attention. The Fig. 2 shows the main causes for traffic accidents.

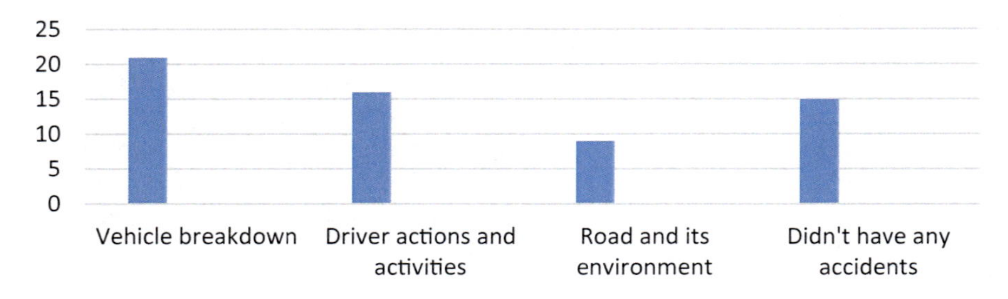

Fig. 2. Main causes for traffic accidents.

A positive result can be regarded the fact that 25% of companies engaged in dangerous cargo transportation, did not have their vehicle units to be involved in traffic accidents. However, in a long run, this percentage may steadily increase due to the significance that the reduction of accidents has on the environment and values.

It is a top priority for Lithuanian economy and its overall image among developed countries to reduce accident numbers involving dangerous cargo. The total number umber of traffic accidents has a tendency to decrease However, it remains relatively high, and efforts should be made, as well as diverse measures should be taken to reduce these statistics and practises.

Figure 3 shows the tendency for carriers to choose a route for the transport of dangerous goods.

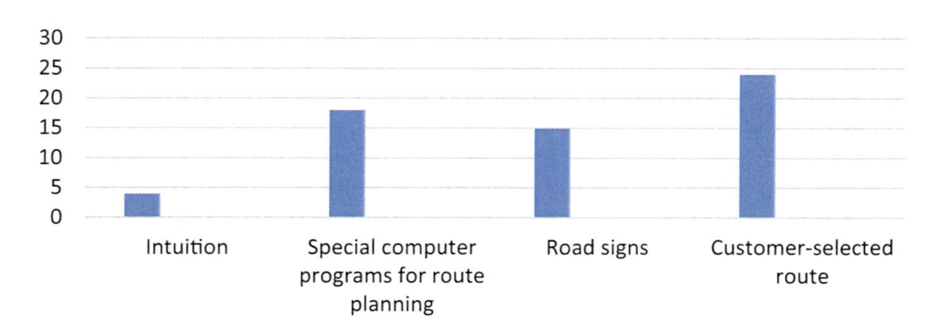

Fig. 3. Factors determining route selection.

Analyzing the data, it can be stated that its Lithuanian carriers pay insufficient attention to route selection. Routes are usually made without taking into account certain hazard factors, and carriers do not use routing methods applied in global practice, little is known about segment selection criteria. This situation shows that carriers' work and research proposals do not work as a unified system.

3 Recommendation Measures for Reducing the Accident Rate

3.1 European Union Safety Measures for the Transport of Dangerous Goods

The EU directive 2008/68/EC [2] has a great impact on reducing the number of accidents whilst transporting dangerous goods. According to this directive, member states of the European Union may even apply stricter provisions in regard to increasing safety measures of dangerous goods' transportation. However, no exceptions shall be made to design requirements for new or already operating vehicles, wagons and inland waterway vessels registered at national territories. Should incidents occur on the territory of a Member State and the Member State considers that the safety measures used are not sufficient to ensure the safe transport of dangerous goods, Member States may inform the Commission on the measures they intend to use to enhance the security of the transport of dangerous goods. The establishment of a single transport market in European Union requires a continuous improvement of preventive measures aimed at ensuring safe organization of dangerous goods. Prior proceeding with transport process of dangerous goods, it is necessary to establish a flawless operation mechanism that eliminates human factor errors in the following processes: packaging, marking, document proceedings and transportation. After all, suitable packaging is one of the main protective measures used for dangerous goods' transportation. Properly packed hazardous materials can protect both – the environment and people involved.

The European safety and health policy cover risk assessment and implementation of prevention measures by giving priority to risk elimination in its occurrence source. These principles should be applied also in the implementation of the safety plan, covering rail vehicles and equipment, activity planning and personnel. The recommendation is to make a record of accidents. Accidents may be analysed as an integral part of risk assessment, contributing to identification of what actions should be best taken.

The number of accidents in transport must be reduced, while this may be done by:

- Reducing the amount of dangerous goods for one transportation unit, as the smaller amount reduces the risk of threat to the public and the environment;
- Ensuring the quality of loading, container, mounting, transporting and unloading;
- Giving attention to climatic conditions, to avoid bad weather conditions, when it is poor visibility level;
- Ensuring preparation, experience and knowledge of the driver and other participants related to transportation while transporting dangerous goods;
- To implement prevention measures, at the moment of which attention would be paid to technical state of vans, ensuring tightness of containers, marking;
- Taking into consideration the incidents that already took place, to take all measures to avoid similar situations;
- Installation of new technologies.

3.2 Lithuanian Republic Government Safety Program

Following the programme of the Government of the Republic of Lithuania for 2008–2025 the Ministry of Transport and Communications is obliged to develop a balanced communication system, reduce the negative of impact of vehicles on the environment and human health. In view of growing volumes of dangerous goods' transportation, and in order to ensure the safety of this type of cargo transport, the Republic of Lithuania has acceded to all transport modes, except for inland waterways, international treaties regulating the transport of dangerous goods (ADR, COTIF, RID and other).

The risk can be reduced by many factors that are closely related:

- Increasing the number of shipments in order to maintain the same amounts of goods transfers; decreasing the goods quantity for one shipment, the total number of those shipments should increase, but it is not an economic solution, and the effect of this risk possibility reduction is fairly equal to the decrease in the probability of accident [5, 6];
- Ensure the quality of the packaging, loading, reloading and fastening of dangerous goods
- Reduction of goods quantity in one shipment, – this is the opposite action from the reduction of an accident possibility, but a smaller amount of dangerous material directly results in reduced level of harm, influence on people and surroundings;
- Correctly chosen route [1].

All risk mitigation conditions can be classified as qualitative and quantitative according to the volume of freight being transported and the frequency of shipment. The risk factor for damage is a qualitative factor since the damage is directly dependent on the hazard of the substance. Accident risk is considered to be a quantitative factor because it is directly dependent on the number of shipments [3, 6].

The unsafe transportation of chemicals and hazardous waste is a major cause for pollution of the groundwater and soil posing a threat to the environment and human health. To avoid this pollution, the best technical measures and economic instruments are applied.

The regulations on waste management cover certain requirements on hazardous waste transportation. The hazardous waste must be packed in a way that would not harm human health or the environment:

- packages or containers must be designed in a way that prevents dangerous waste from spillage, spreading, evaporating or otherwise get into the environment;
- packaging materials must be resistant to the effects of hazardous waste and its components; and do not manifest any reactions to these wastes and its components;
- packaging and containers, and its components must be firm, tight and hermetic, so as not to break, loosen up, or open up during transportation process and materials within are not released into the environment;
- containers with reusable covers and plugs must be manufactured and designed in a way that can be safely opened and closed and whilst these processes, waste and other components are not released into the environment;
- all containers and packaging that are securely transported must have a special marking.

4 Conclusions

1. Transportation of dangerous goods is one of the most complex and mostly safety requiring transportation technologies. Because of peculiarities and risk, transportation must be precisely controlled, regulated and handled.
2. Based on the data of the research, enterprises should pay attention to measures helping to prevent accidents as well as measures contributing to reduction of accident consequences.
3. Assessing the risk creates the possibility for carriers to choose basic transport criteria and flexibility. Risk assessment allows reducing the likelihood of accidents and increasing transport safety.
4. Analyzing the data, it can be stated that its Lithuanian carriers pay insufficient attention to route selection. Routes are usually made without taking into account certain hazard factors, and carriers do not use routing methods applied in global practice, little is known about segment selection criteria. This situation shows that carriers' work and research proposals do not work as a unified system.

References

1. Batarlienė, N.: Risk and damage assessment for transportation of dangerous freight. Transp. Telecommun. J. **19**(4), pp. 356–363 (2018). Warsaw; Riga: De Gruyter Open; Transport and Telecommunication Institute. ISSN 1407-6160
2. Blanco, M.A.: Safety adviser for the transport of dangerous goods by road, Seguridad sy medio ambiente Nr. 123 (2011). http://www.mapfre.com/fundacion/html/evistas/seguridad/n123/docs/Articulo4en.pdf. Accessed 28 Dec 2018
3. Directive 2008/68/EC of the European Parliament and of the Council of 24 September 2008 on the inland transport of dangerous goods. Official Journal of the European Union, L 260/13, 30 September 2008
4. Janno, J., Koppel, O.: Managing dangerous goods risks on roads during transportation under normal conditions. DAAM Int. Sci. Book **25**, 333–344 (2017)
5. Ranitović, P., Tepić, G., Sremac, S.: Synergy between ISO 14001 and the criteria for hazardous material transport. In: Management of Technology Step to Sustainable Production, MOTSP 2011, pp. 48–53 (2011)
6. Najib, M., Boukachour, H., Boukachour, J.: Multi-agent framework for hazardous goods transport risk management (2009). http://www.srlst.com/ijist/special%20issue/ijism-special-issue2010-2_files/Special-Issue2010_2_27.pdf. Accessed 16 Jan 2019
7. Thomson, J.B.: International co-operation in hazardous materials accident prevention. J. Loss Prev. Process Ind. **12**, 217–225 (1999)
8. Tomasoni, M.A., Garbolino, E., et al.: Risk evaluation of real-time accident scenarios in the transport of hazardous material on road. Manage. Environ. **21**(5), 695–711 (2010). https://doi.org/10.1108/14777831011067962
9. UNECE Homepage. http://www.unece.org. Accessed 02 Apr 2019

A Geospatial Multi-scale Level Analysis of the Distribution of Animal-Vehicle Collisions on Polish Highways and National Roads

Rob Smits[1], Janusz Bohatkiewicz[2], Joanna Bohatkiewicz[3(✉)], and Maciej Hałucha[1]

[1] EKKOM Sp. z o.o., ul. dr., Józefa Babińskiego 71B, 30-394 Kraków, Poland
[2] Department of Roads and Bridges, Lublin University of Technology, ul. Nadbystrzycka 40, 20-618 Lublin, Poland
[3] Department of Management and Social Communication, Jagiellonian University, ul. prof. Stanisława Łojasiewicza 4, 30-348 Kraków, Poland
joanna.bohatkiewicz@gmail.com

Abstract. This article has the aim to investigate the spatial distribution of animal-vehicle collisions (AVCs) in Poland on national roads and motorways. Besides identifying hot spot locations of animal mortality within the country, it also researches on regional scale the relationship between levels of traffic volume or traffic speed and the amount of AVCs. The analysis involves data collected by the Polish General Directorate for National Roads and Motorways with 2014 until 2017 as time range. The geospatial analyses are carried out with QGIS. The Małopolska Province has the highest concentration of AVCs in Poland. Results of the analysis on regional level show that traffic speed and traffic volume are both not strongly correlating to the total animal mortality, meaning that higher levels of the parameters do not lead to higher amounts of AVCs. Accidents occur most frequently on road sections with a traffic volume of an average number of 10,000 to 15,000 vehicles per day. The severe accidents with large sized animals, the Big-Four, are occurring most often at road sections with speed limits between 70 and 90 km/h. This work will be continued, which will investigate the influence of the landscape adjacent to the road on a local scale.

Keywords: Animal-vehicle collisions · Animal mortality · Animal passages

1 Introduction

The vehicle is an essential aspect of the modern-day life of millions of people around the world. Unfortunately, this necessity has its consequences. Animals are getting directly and indirectly confronted in several ways by roadway traffic. Direct effects are mainly the animal-vehicle collisions, decreased habitat amount, fragmented populations and resource inaccessibility [9, 12, 16]. An example of the degree of the direct impact of roadways on fauna is the estimation of Conover et al. [5], stating that approximately 1.5 million collisions occur annually between deer and automobiles in

A. Varhelyi et al. (Eds.): VISZERO 2018, LNITI, pp. 74–84, 2020.
https://doi.org/10.1007/978-3-030-22375-5_9

the United States alone. In some cases the victims are not only the animals, Brieger et al. [3] mentions yearly an estimation of 500 human fatalities in the United States and Europe resulting from the collisions with heavy bodied animals on roads.

Roads and cars affect animals also in other direct ways, such as acoustic emission, chemical emission, and light emission of cars [1, 14]. This article, however, focuses only on the direct impact of roads on the animals, namely the animal-vehicle collisions (AVCs) and the spatial distribution of the animal mortality on roads in the Polish context. Is the traffic speed or the traffic volume parameter increasing or decreasing the probability of an AVC? This question will be pivotal in this article.

2 Impact of Road Traffic on Animal Mortality

To analyse the spatial distribution of AVCs and the potential reasons for that spatial distribution, parameters are selected based on their previously found relationship to AVCs. The parameters are grouped in the following categories: vehicle and road; environmental; and animal parameters.

Vehicle and Road Parameters. The traffic volume parameter is frequently included in predictive models related to animal mortality on roads [9, 21]. Seiler and Helldin [18] mention that low traffic volume in combination with high speed levels results in the highest animal mortality rates, due to the long time gaps between vehicles, stimulating an animal to attempt to cross the road and getting collided with a rapidly approaching car. Seiler [17] predicted the existence of the so called "deadly trap" meaning that more than 50% of the animals attempting to cross the road get killed when a road section has between 2,500 and 10,000 vehicles on an average day, see Fig. 1. This is a result of a predictive model and is an indication of the level of traffic volume where the total barrier starts. The results chapter will refer to this figure and be compared with the analysed data in the study area of this article.

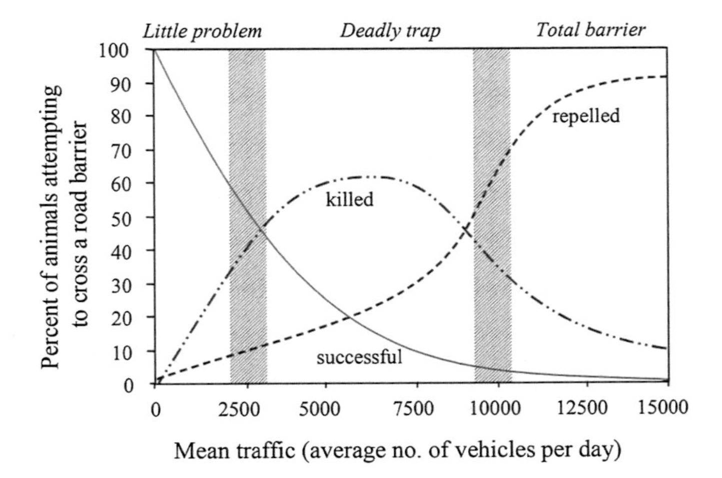

Fig. 1. Conceptual model of Seiler showing the percent of animals that successfully cross, are repelled, or get killed as they attempt to cross the road versus traffic volume [17].

Environmental Parameters. Dodd et al. [6] confirm that road barriers, such as fencing, have impact on the amount of AVCs, although it can also increase the barrier effect of a road network blocking animal movement [8]. Wildlife crossing structures combined with fencing can help increase the sustainability of animal populations and many scientific researches try to measure the effect of wildlife crossing structures on AVCs [1, 2, 4, 19]. The amount of AVCs on a road section surrounded by various land use categories is also often examined like: forest, rural, and built-up areas adjacent to roads [11, 13, 18, 20]. Certain areas of land are, due to the frequent movement of animals, mentioned as conservation corridors and the vicinity of a road to corridors is in AVC related research also used as a parameter [10].

Animal Parameters. Animals getting collided can differ in kind of species, size, weight and behavioural responses. Especially the size and weight of an animal can influence the impact of an AVC on the human. Large, heavy-weight animals such as moose (Alces alces) gets often involved in AVCs possibly resulting in human fatalities [7].

3 Method of Research and Analysis

Study Area. This article starts with the spatial distribution of the AVC data on national scale, which is the country of Poland. Subsequently, the Province that reveals the highest frequency of AVCs is selected as the area of interest for the regional scale analysis.

Scope. This article tries to identify the potential relationship between the vehicle and road parameters, traffic volume and traffic speed, and spatial distribution of AVCs. The environmental parameters, although proven in the literature to have an impact on AVCs, will be left out of the analysis. This is decided due to the complexity of the environmental parameters and it will be discussed separately in the follow-up research. To take into account the difference between species involved in AVCs, specific attention will be paid to large-sized mammals that frequently cause severe accidents, namely roe deer (Capreolus), red deer (Cervus), moose (Alces alces) and wild boar (Sus scrofa). These four kinds of animals are called "Big Four" in the study.

Data. The AVC, vehicle and road parameter data used in this article are collected by the Polish General Directorate for National Roads and Motorways (Generalna Dyrekcja Dróg Krajowych i Autostrad in Polish). As of now this body will be referred to as GDDKiA. The road sections where data is collected are not only highways but also national roads of importance within the regional infrastructure. The AVC data of 2014–2016 consists of a total amount of 46,366 observations and will be used for the national scale. The dataset contains AVC involved species ranging from birds, domestic animals, large and small-sized mammals. Due to the availability of data, the analysis the AVC observations for the year 2017 are added on regional level. Lastly, the traffic volume dataset represent the average total number of passenger cars and trucks in 24 h.

Software and GIS Method. The GIS software that will be used for analyzing the spatial distribution of the animal mortality data is the open source QGIS. In order to examine the data, the spatial analytic 'Heat map' tool is applied [15].

4 Results

National Scale. Figure 2 is the first overview of the spatial distribution of the animal mortality in Poland. The AVC value presented in the map is the result of the Kernel density estimation of AVCs within the radius, or Kernel bandwidth, in meters [15]. In Poland three Provinces have high frequency of AVCs, namely: Kujawsko-Pomorskie, Swiętokrzyskie and Małopolskie Provinces. Małopolska in the south of the country, has the largest concentration of AVCs and is selected as the area of interest for the regional scale analysis involving the road and vehicle parameters.

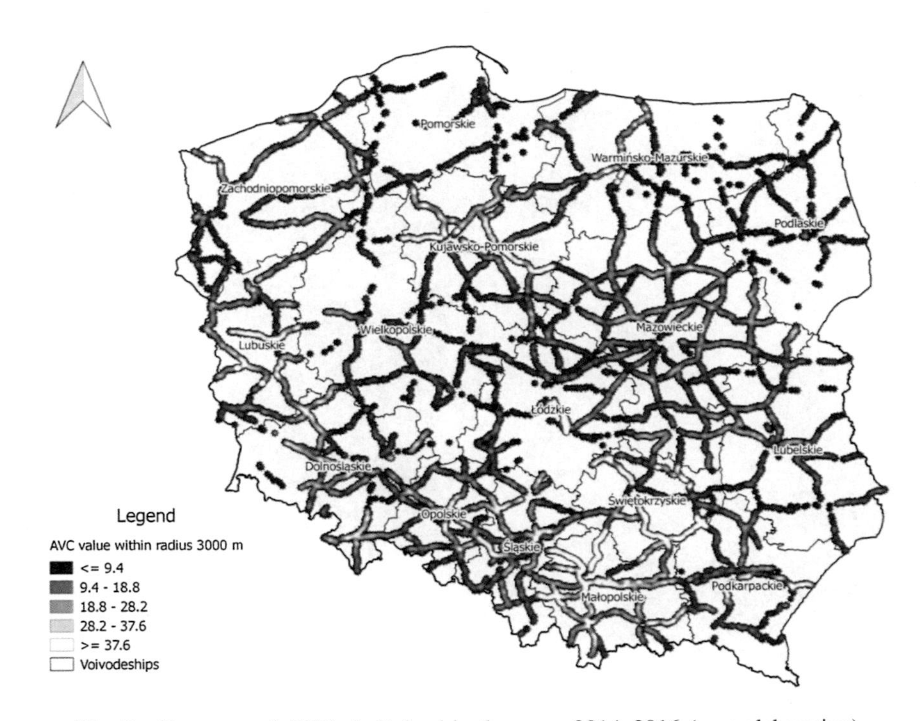

Fig. 2. Heat map of AVCs in Poland in the years 2014–2016 (own elaboration).

Regional Scale. Figure 3 visualizes the spatial distribution of both the total amount of AVCs and Fig. 4 presents the accidents involving the Big-Four in the Małopolska Province. The map on top shows a large concentration of AVCs at the roads north and north-west of the city of Krakow. In addition, the route from Krakow, leading to the south has also very high values. While focusing on the map of the accidents with

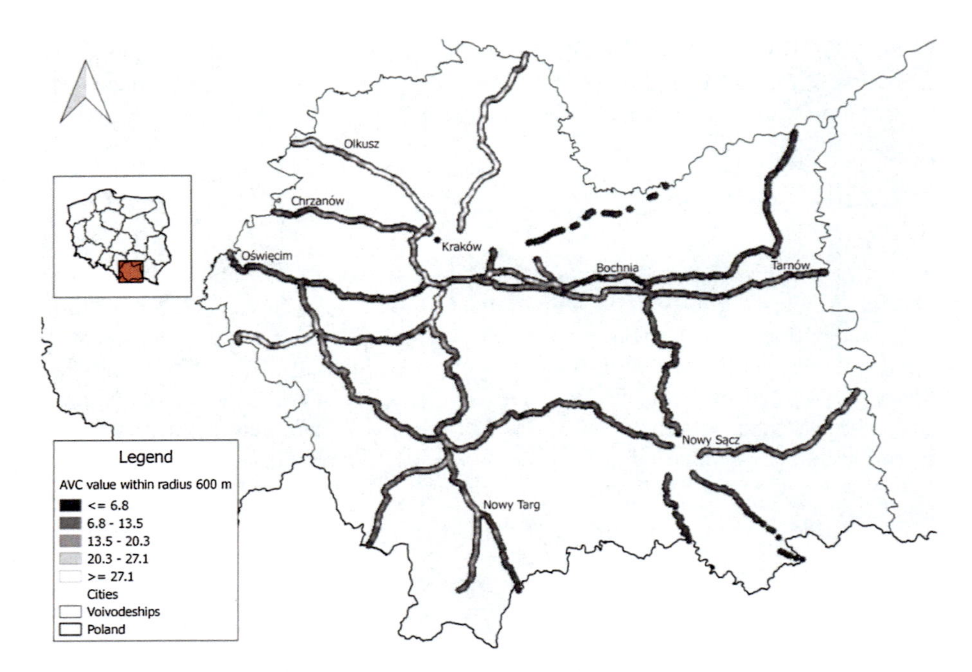

Fig. 3. Heat map of AVCs in Małopolska in the years 2014–2017 (own elaboration).

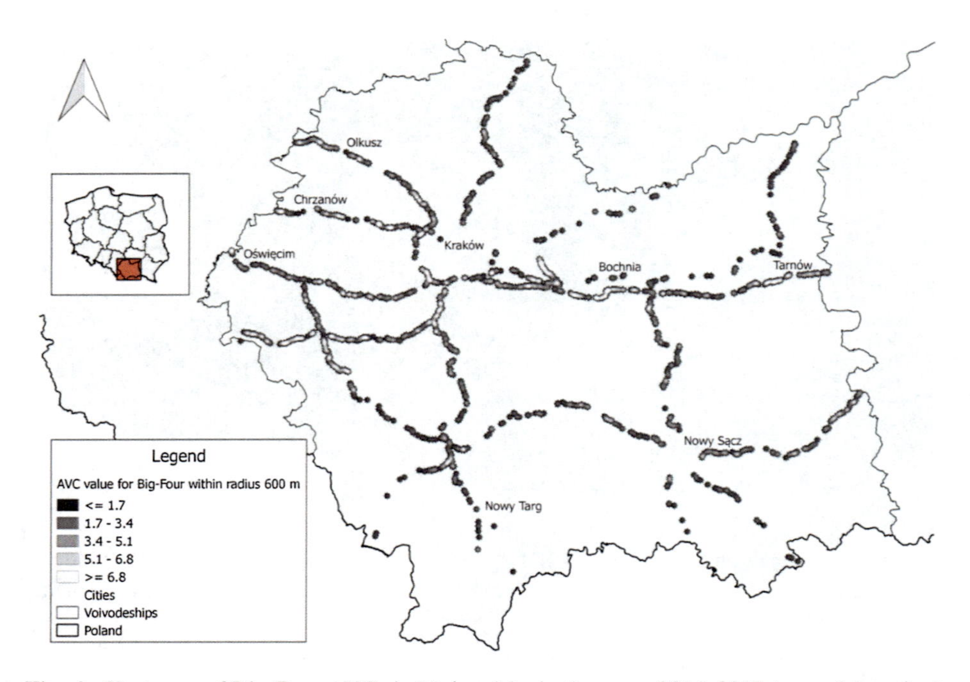

Fig. 4. Heat map of Big-Four AVCs in Małopolska in the years 2014–2017 (own elaboration).

the Big-Four, it turns out that the high density of accidents in the north of Krakow has disappeared. The highest concentration of accidents with the Big-Four in the Province is now on the roads east of Krakow, leading to Bochnia and Tarnow.

Traffic Volume. Figure 5 shows that almost 50% of all accidents between 2014 and 2017 in Małopolska occurred at a traffic volume of 10,000 to 15,000 vehicles per day. The space between the traffic volume levels can be named the 'deadly trap' for the Małopolska study area.

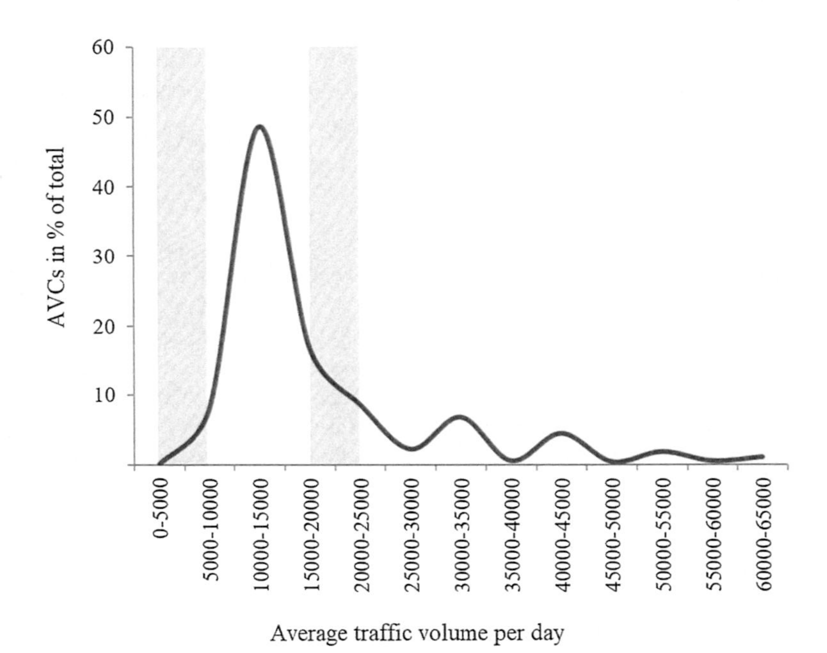

Fig. 5. Distribution of the total AVC data of the years 2014–2017 in percentage of the total observations (13,641) in Małopolska related to the average daily traffic volume (own elaboration).

In Fig. 6 the number of AVCs is aggregated per kilometre of road with a certain traffic volume. This is done both for the total amount of AVC and the accidents with the Big-Four within 2014 and 2017 in Małopolska. The scatter plot reveals that there is a moderately positive correlation between the total amount of AVCs and high traffic volume levels. However, the correlation is not strong and therefore the statement can be made that higher traffic volumes do not lead to higher rates of AVCs per kilometre of road. The correlation between accidents with the Big-Four per kilometre and the traffic volume is also moderately positive, but less pronounced than the relationship between the total number of accidents and the traffic volume. Also to the Big-Four AVCs applies, higher traffic volume does not per definition lead to a higher number of accidents per kilometre.

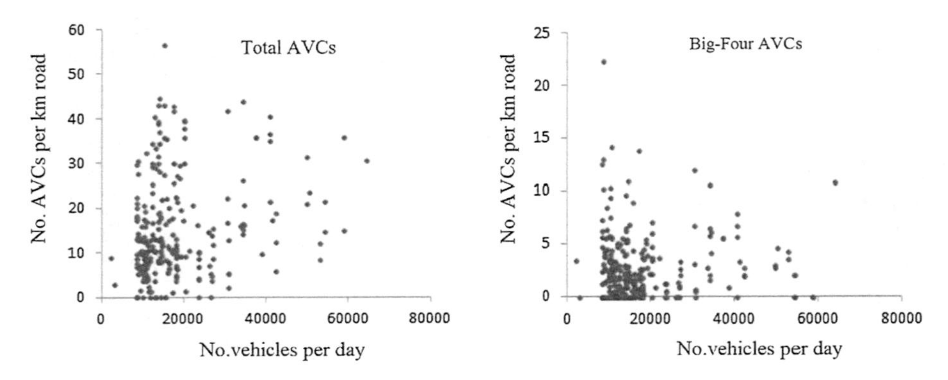

Fig. 6. Scatter plot with the average daily traffic volume and the number of AVCs per kilometre of road in the years 2014–2017 in Małopolska (the total number of AVCs at the left and the Big-Four AVCs at the right) (own elaboration).

Figure 7 presents the total amount of AVCs related to the maximum speed limits of the road where the AVC occurred. As it can be seen in Fig. 7, the roads with speed limits below 100 km/h are predominantly the roads where AVCs occur.

In Fig. 8, the number of accidents is standardised again by the number of accidents per kilometre of road and is shown in scatter plots. For both the total AVC and the Big Four AVCs it can be stated that a road with higher maximum speed limit does not lead to higher frequency of accidents with animals per kilometre of road. In the scatter plot of the total number of AVCs, the impact rate per kilometre is the highest on roads with

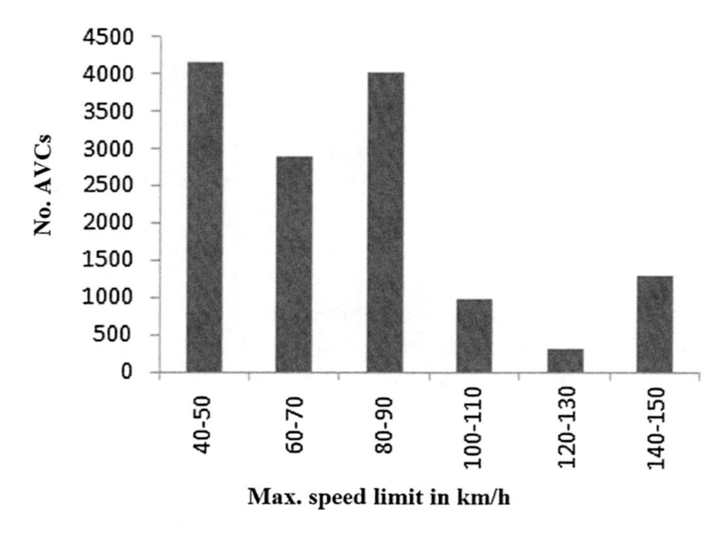

Fig. 7. Distribution of the total AVC data observations (13641) of the years 2014–2017 in Małopolska related to the traffic speed (maximum speed limit in kilometre per hour) (own elaboration).

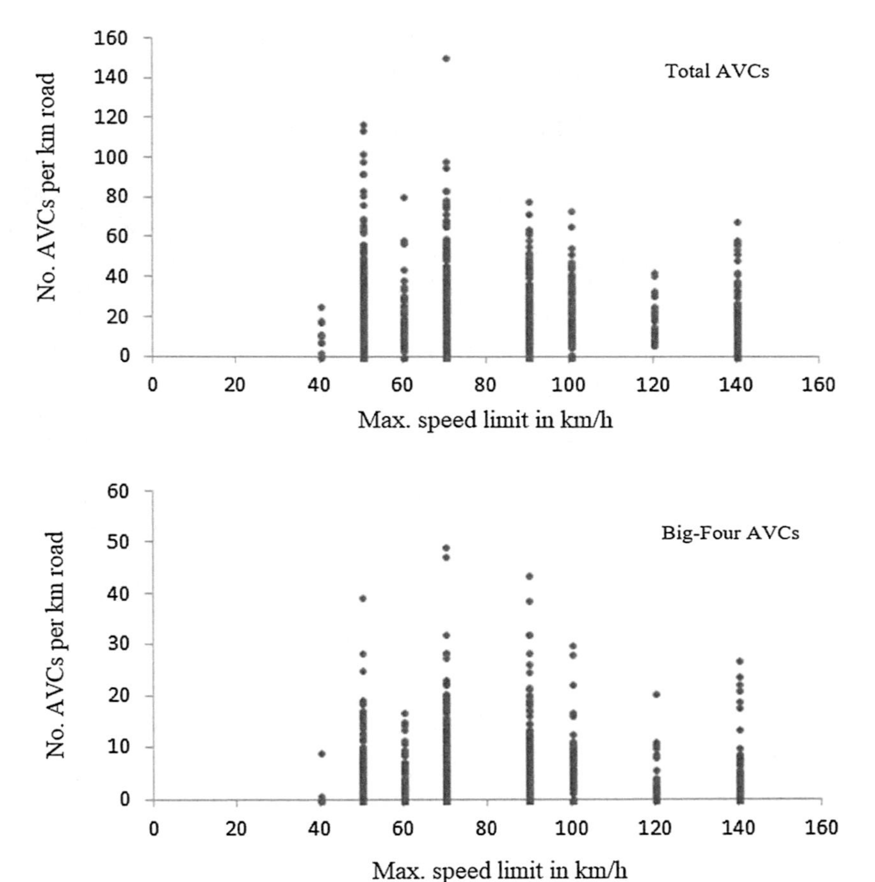

Fig. 8. Scatter plot with the maximum speed limit of a road in kilometre per hour and the number of AVCs per kilometre of road in the years 2014–2017 in Małopolska (the total number of AVCs on top and the Big-Four AVCs below) (own elaboration).

maximum speed limits of 50 to 70 km/h. The accidents between vehicles and the Big-Four animals occur with the highest frequency per kilometre on roads with maximum speed limits between 70 and 90 km/h.

5 Discussion

The roadways have a direct impact on fauna due to animal-vehicle collisions (AVCs), causing an estimated 500 human fatalities in the United States and Europe annually [3]. The aim of the article is to find out to what extent the vehicle and road parameters have an influence on the spatial distribution of the animal mortality on the road. The AVC data has been gathered by the Polish General Directorate for National Roads and Motorways (GDDKiA) between 2014 and 2017. At the national scale, the results of the geospatial analyses identify the Małopolska Province as an area with a relatively high density of AVCs compared to the rest of Poland. On the regional scale that focuses on Małopolska, the concentration of AVCs is located at the roads at the north and northwest of the city of Krakow. The most severe accidents involving the Big-Four animals, namely: roe deer (Capreolus), red deer (Cervus), moose (Alces alces) and wild boar (Sus scrofa), occur on the route that leans east from Krakow to Bochnia and Tarnow. Statistical analysis finds out that both traffic volume and traffic speed parameter are not strongly correlating with the total amount of AVCs and Big-Four AVCs per kilometre. The probability of an AVC is higher with low and medium levels of traffic volume. This is in line with the predications made by Seiler [17]. The reason for this is expected to result from the avoidance behaviour of animals on a road with such a high level of traffic volume that it becomes a total barrier [17]. Any attempt to cross a road with a high traffic volume level causes a collision with a vehicle. The data for Małopolska shows the existence of the deadly trap and the traffic volume level causing a total barrier for animals, as predicted by Seiler [17]. However, the total barrier starts at a traffic volume level of approximately 17,500 vehicles per day and not at 10,000. The total amount of AVCs occur the most often on roads with speed limits of 50 to 70 km/h. The Big-Four AVCs happen most frequently on roads with maximum speed limits of 70 to 90 km/h. On hot spot locations of AVCs with the Big-Four follow-up research has to be carried out, taking into account the environmental parameters in order to increase the safety for both animals as well as drivers.

Limitations and Recommendations. The article does not specifically analyse all different kinds of species involved in collisions with vehicles, such as domestic animals and birds. The reason for analysing the Big-Four AVCs is discussed in the scope section. The dataset used in the article is extensive but still limited to national roads selected by the GDDKiA. This means that not all national roads in Poland are included. Moreover, the road and vehicle parameters are not the only factors that may influence the probability of AVCs. Another option is the environmental setting on AVC hot spot locations, such as the surrounding land use, the ecological corridor areas and the applied mitigation methods [11, 13, 18, 20]. This research has confirmed that roads with high traffic volume generate a strong barrier effect of animals getting repelled or with near certainty collided while trying to cross. The mitigation methods, such as overpasses (animal passageways) should be provided for the roads on these areas in order to avoid severe accidents and decrease the barrier effect. On the local scale, it is necessary to investigate what other factors cause the hot spots of AVCs shown on the heat maps. General observation on the local scale indicates that the planning of a construction of mitigation methods such as overpasses should cover a larger focus area, not only highways. National roads seem to be the roads with high concentrations of accidents, especially with the Big-Four animals. Whether this truly is the case, will be researched using environmental analysis involving GIS and presented in the follow-up article.

References

1. Beben, D.: Crossings construction as a method of animal conservation. Transp. Res. Procedia **14**, 474–483 (2016)
2. Bliss-Ketchum, L.L., de Rivera, C.E., Turner, B.C., Weisbaum, D.M.: The effect of artificial light on wildlife use of a passage structure. Biol. Conserv. **199**, 25–28 (2016)
3. Brieger, F., Hagen, R., Vetter, D., Dormann, C.F., Storch, I.: Effectiveness of light-reflecting devices: a systematic reanalysis of animal-vehicle collision data. Accid. Anal. Prev. **97**, 242–260 (2016)
4. Clevenger, A.P., Ford, A.T.: Wildlife crossing structures, fencing, and other highway design considerations. Safe Passages–Highways, Wildlife, and Habitat Connectivity, p. 424 (2010)
5. Conover, M.R., Pitt, W.C., Kessler, K.K., DuBow, T.J., Sanborn, W.A.: Review of human injuries, illnesses, and economic losses caused by wildlife in the United States. Wildl. Soc. Bull. (1973–2006) **23**(3), 407–414 (1995)
6. Dodd Jr., C.K., Barichivich, W.J., Smith, L.L.: Effectiveness of a barrier wall and culverts in reducing wildlife mortality on a heavily traveled highway in Florida. Biol. Conserv. **118**(5), 619–631 (2004)
7. Eid, H.O., Abu-Zidan, F.M.: Biomechanics of road traffic collision injuries: a clinician's perspective. Singapore Med. J. **48**(7), 693 (2007)
8. Jaeger, J.A., Fahrig, L.: Effects of road fencing on population persistence. Conserv. Biol. **18** (6), 1651–1657 (2004)
9. Jaeger, J.A., Bowman, J., Brennan, J., Fahrig, L., Bert, D., Bouchard, J., von Toschanowitz, K.T.: Predicting when animal populations are at risk from roads: an interactive model of road avoidance behavior. Ecol. Model. **185**(2–4), 329–348 (2005)
10. Garrah, E., Danby, R.K., Eberhardt, E., Cunnington, G.M., Mitchell, S.: Hot spots and hot times: wildlife road mortality in a regional conservation corridor. Environ. Manage. **56**(4), 874–889 (2015)
11. Gunther, K.A., Biel, M.J., Robison, H.L.: Factors influencing the frequency of road-killed wildlife in Yellowstone National Park. In: International Conference on Wildlife Ecology and Transportation (ICOWET 1998) Florida Department of Transportation US Department of Transportation US Forest Service Defenders of Wildlife (1998)
12. Lima, S.L., Blackwell, B.F., DeVault, T.L., Fernández-Juricic, E.: Animal reactions to oncoming vehicles: a conceptual review. Biol. Rev. **90**(1), 60–76 (2015)
13. Orlowski, G., Nowak, L.: Road mortality of hedgehogs Erinaceus spp. in farmland in Lower Silesia (south-western Poland). Pol. J. Ecol. **52**(3), 377–382 (2004)
14. Proppe, D.S., McMillan, N., Congdon, J.V., Sturdy, C.B.: Mitigating road impacts on animals through learning principles. Anim. Cogn. **20**(1), 19–31 (2017)
15. QGIS: Heatmap Plugin. https://docs.qgis.org/2.18/en/docs/user_manual/plugins/plugins_heatmap.html. Last accessed 14 Jan 2019
16. Rytwinski, T., Fahrig, L.: The impacts of roads and traffic on terrestrial animal populations. In: Handbook of Road Ecology, pp. 237–246 (2015)
17. Seiler, A.: The toll of the automobile: wildlife and roads in Sweden. Dissertation, Swedish University of Agricultural Sciences, Uppsala (2003). https://pub.epsilon.slu.se/388/1/Silvestria295.pdf. Last accessed 14 Jan 2019
18. Seiler, A., Helldin, J.O.: Mortality in wildlife due to transportation. In: The Ecology of Transportation: Managing Mobility for the Environment, pp. 165–189. Springer, Dordrecht (2006)

19. Smith, D.J., van der Ree, D., Rosell, C.: Wildlife crossing structures: an effective strategy to restore or maintain wildlife connectivity across roads. In: Handbook of Road Ecology, pp. 172–183 (2015)
20. van Langevelde, F., Jaarsma, C.F.: Modeling the effect of traffic calming on local animal population persistence. Ecol. Soc. **14**(2), 39 (2009)
21. van Strien, M.J., Grêt-Regamey, A.: How is habitat connectivity affected by settlement and road network configurations? Results from simulating coupled habitat and human networks. Ecol. Model. **342**, 186–198 (2016)

Impact of Road Traffic Accidents on the Dynamics of Traffic Flows

Algimantas Danilevičius(✉) and Marijonas Bogdevičius

Department of Mobile Machinery and Railway Transport, Vilnius Gediminas Technical University, Plytinės Str. 27, 10105 Vilnius, Lithuania
{algimantas.danilevicius, marijonas.bogdevicius}@vgtu.lt

Abstract. Traffic flows can be affected by any changes on the road. Particularly large changes can be caused by traffic accident. Using the discrete traffic flow method, the dynamics of intensive traffic flow on the road was analyzed. As a research object chosen one of intensive road in Vilnius city. There are eight traffic lights in the intensive road, with a 120 s period cycle. On this road, one traffic accident is simulated on a certain road section, which occur at same times, but different traffic accident elimination time. The changes in traffic flow parameters are obtained at different traffic accidents, which almost completely stops the traffic flow, and their removal times. The results of the research are presented as the parameters changes in traffic flow in time and characteristic waypoints. The dynamics of traffic flows is investigated, after eliminating the consequences of the traffic accidents. Setting the time at which the traffic flows up to the start of traffic accidents.

Keywords: Discrete method · Traffic flow · Traffic light · Traffic accident · Dynamic

1 Introduction

To better understand how traffic flows respond to different situations on the road, different research methods are being developed. Micro and macro traffic flow models are used to simulate traffic flows in more detail. In article [1] authors developed a route-based flow model to explore the effects of bus station and accident on network traffic flow. The numerical results show that the proposed model can qualitatively reproduce some complex traffic phenomena caused by these factors in a network (e.g., shock, rarefaction wave, jam, stop-and-go, etc.). The authors used the Courant-Friedrichs-Lewy condition for better results. Authors in article [2] analysis safety investigating four types of paths in a freeway, namely two straight lanes, three straight lanes, ramps, and roundabouts as case studies and discuss the different traffic rules as comparison. For that analysis authors used Cellular automata traffic flow model. Bottleneck (especially the moving bottleneck) widely exists in the urban traffic system [3]. Authors incorporate the propagation speed of a moving bottleneck into the traffic flow model, and then develop an extended macro model to study the impacts of a moving

A. Varhelyi et al. (Eds.): VISZERO 2018, LNITI, pp. 85–92, 2020.
https://doi.org/10.1007/978-3-030-22375-5_10

bottleneck on traffic flow under two typical traffic situations. The numerical results indicate that the influences are directly dependent on the initial traffic density. Authors [4] state an optimal control problem using the maximal speed of the coordinated vehicle as control variable. They use the Model Predictive Control approach to get a fuel consumption reduction when the traffic is congested due to the presence of a fixed downstream bottleneck on the highway. In addition they show that the control application also improves the average travel time and queue length. Using Lighthill–Whitham–Richards model authors develop an extended macro model for traffic flow with consideration of multi static bottlenecks to study the effects of the number of static bottlenecks and the spatial distance between two adjacent static bottlenecks on traffic flow [5]. Authors [6] using Nagel–Schreckenberg model also investigate 3 types of bottleneck expansions: serial, parallel, combination. Analysis shows that for largescale expansion, a new class of processing bottleneck known as the serial bottleneck is more efficient than the conventional parallel bottleneck in the absence of human driving behaviour. Traffic instability is better characterized by traffic speed than density [7]. Paper [8] quantitatively investigated the relationship between traffic bottleneck roads and the related intersection signal timings. With adopting ant colony algorithm seeking for the optimal signal timing plan for a regional network an optimization method was proposed with aim to mitigate the bottleneck degree through reducing the risk of high variance in bottleneck indicators.

In order to determine how traffic flows are affected by unexpected obstacles on the road, a goal has been set – to determine the changes of traffic flow dynamics in the event of traffic accident and after it eliminated. The following tasks have been accomplished for this purpose:

- to create a road section dynamic model with traffic lights;
- to choose the boundary and initial conditions of traffic flow variables;
- to determine the influence for traffic flow according to the duration of the elimination of the traffic accident;
- to determine the time to recover the traffic flow to the state that was before the accident.

2 Research Method

The discrete traffic flow method was used to analyze traffic flows. Using this method, it is possible to analyze the variation of traffic flow parameters in time at any point of the modeled road (Fig. 1). Modeling traffic flow points are connected by the equation system (1) which describes the velocity and density changes in time according to the adjacent simulated waypoints settings. The method is further described in the publications [9, 10].

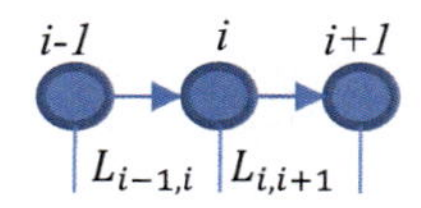

Fig. 1. Calculation scheme which indicates the relationship between the points to be modeled.

$$\begin{Bmatrix} \dot{v}_i \\ \dot{k}_i \end{Bmatrix} = \begin{Bmatrix} F_{v_i}(v_{i-1},\ v_i,\ v_{i+1},\ k_{i-1},\ k_i,\ k_{i+1}) \\ F_{k_i}(v_{i-1},\ v_i,\ v_{i+1},\ k_{i-1},\ k_i,\ k_{i+1}) \end{Bmatrix}, \tag{1}$$

where F_{vi}, F_{ki} – the right side of the vector elements; v_i – velocity of traffic flow at i-point, m/s; k_i – density of traffic flow at i-point, aut./m.

3 Research Object

As a research object chosen one of intensive road – Kalvarijų street in Vilnius city (Fig. 2). The simulated one-way road is divided into sections each L = 50 m. The length of the modeled road is 4.1 km, so all sections are connected by 82 points i. Eight points *i (27, 38, 41, 53, 58, 62, 69, 72)* in the modeled road, it is distinguished by the fact that they are marked with places, where there are intersections with controlled traffic lights.

Fig. 2. Research object: Kalvarijų street, Vilnius (from Ateities street to Žalgiris street direction).

In order to analyze the parameters of traffic flows, a calculation scheme was created (Fig. 3).

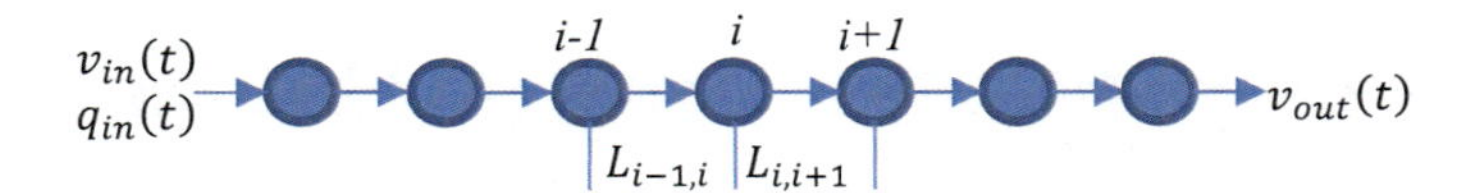

Fig. 3. One-way road section calculation scheme.

4 Research Conditions

4.1 Boundary and Initial Conditions

On the simulated road, all traffic lights cycles are the same. This means that at the same time in all the traffic lights at controlled intersections burn green or red signal. The traffic light period of a full cycle was selected T = 120 s for traffic flow research. The signals period T = 120 s consists of 4 s yellow, 56 s green, 4 s yellow and 56 s red.

In the simulated one-way road, traffic flow parameters are entered with variable flow and velocity according to the Eqs. (2, 3) created by the authors of this article:

$$q_{in}(t) = q_{in0} + A_q sin(w_q t) \tag{2}$$

$$v_{in}(t) = v_{in0} + A_v sin(w_v t) \tag{3}$$

where q_{in0}, v_{in0} – average value of flow [veh./s] and velocity [m/s]; A_q, A_v – amplitude of flow, veh./s and velocity [m/s]; w_q, w_v – angular flow and velocity, [rad/s].

Traffic flows were simulated setting traffic lights cycle to: T = 120 s.

At the last point of the modeled road, the velocity $v(t)$ depends on the current density $k(t)$.

$$v_{out}(t) = \left(1 - \left(\frac{k(t)}{k_{max}}\right)^{a_2}\right)^{a_1} \tag{4}$$

where $k(t)$ – current traffic density at point [veh./m]; k_{max} – maximal traffic density [veh./m].

4.2 Location and Conditions of the Traffic Accident

Traffic accidents are simulated on a 58th road section (Fig. 4), which almost completely stops the traffic flow.

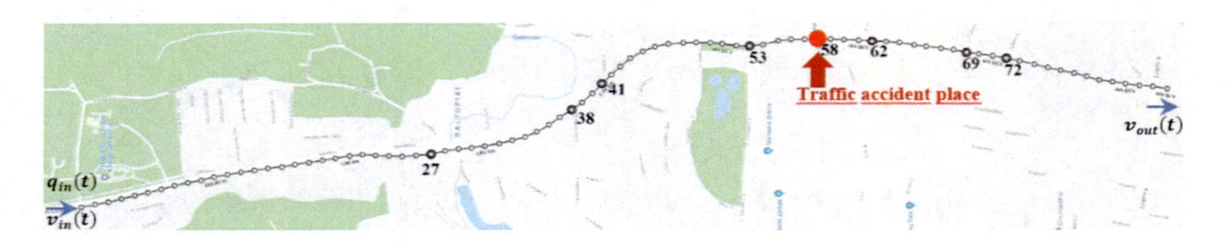

Fig. 4. The location of the traffic accident.

Traffic accidents elimination times are t = [5, 10, 20, 30] min in simulated road. And that traffic accident occurs in the 3,700th simulation second.

5 Results

The results show that the traffic flow parameters (velocity, density, flow) before the traffic accident changes to a steady rhythm (Fig. 5) indicated by the traffic light cycle.

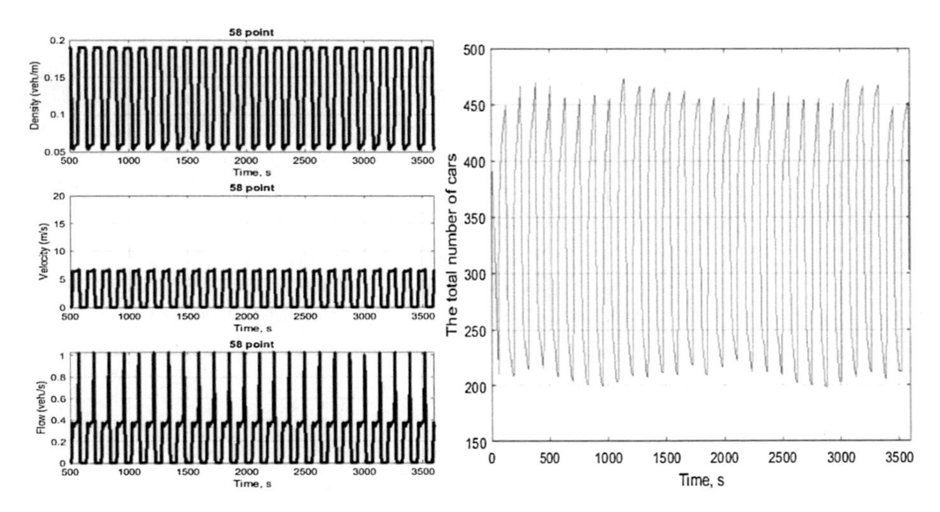

Fig. 5. Results without traffic accidents.

There is an obstacle on the road when an accident occurs at point 58 of the modeled road. Because of this obstacle, road capacity is very low. Before to the obstacle, the density increases significantly, and the velocity at the green signal is very low. As a result, obtained a low flow (Fig. 6).

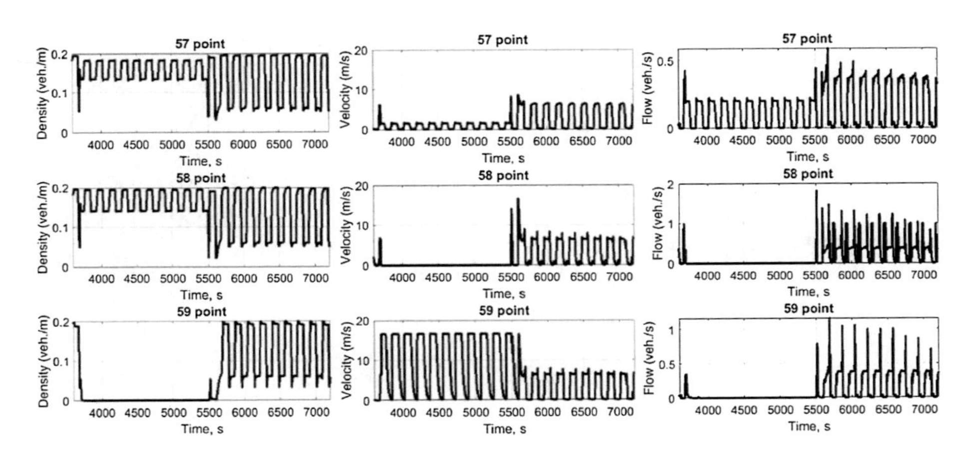

Fig. 6. Results when traffic accidents removal time is 30 min.

The results show that at the intersection where is the traffic accident, the velocity is almost equal to 0 during the green signal, because the vehicles passes the obstacle very slowly. After passing the obstacle, the vehicles can accelerate to the maximum permitted velocity, because the density on the road after traffic accident place is almost equal to 0. But the flow remains very low, because the passing obstacle vehicles number is very small.

Looking at the results showing the total number of cars on the road (Fig. 7), it is noticed that the number of cars suddenly decreases when a traffic accident occurs. This is because it clears the road section behind the traffic accident place, leaving vehicles from simulated road. But at the same time, the density in the section opposite the location of the traffic accident increases, because vehicles periodically enter to the modeled road. The total number of cars is therefore constantly increasing. Depending on the time of the traffic accident elimination time, the total number of cars may be higher than before the traffic accident, although the section after the obstacle is almost empty.

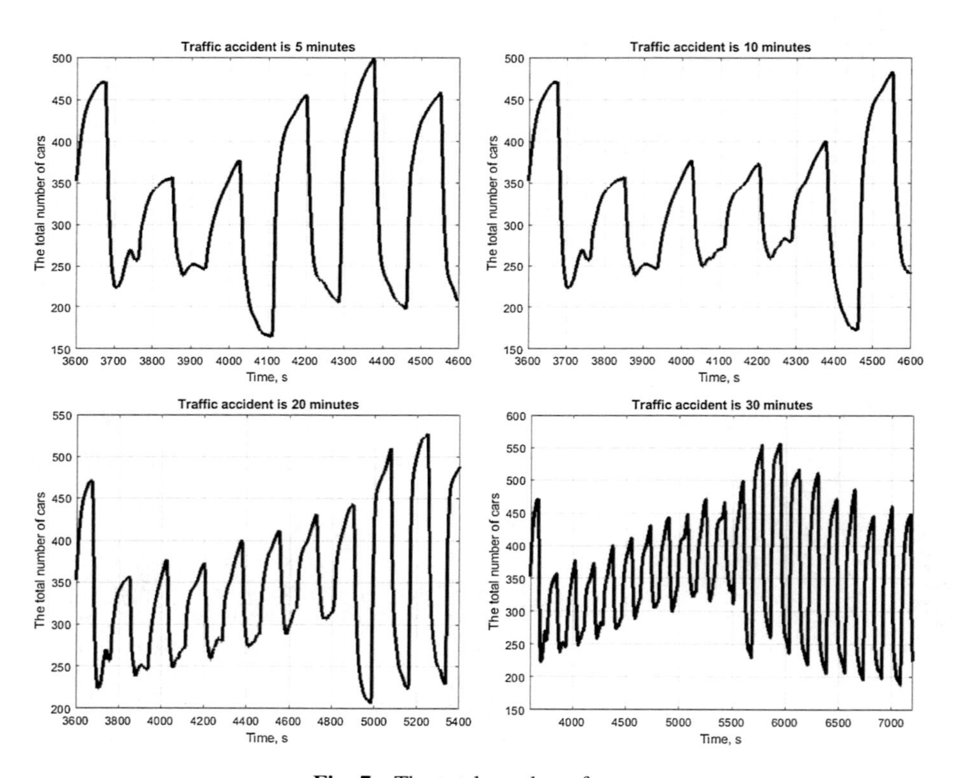

Fig. 7. The total number of cars.

Summarizing the results showing the total number of cars in the event of traffic accident it is observed that there is a linear relationship between the total number of cars (Fig. 8) on the road and traffic accident elimination time:

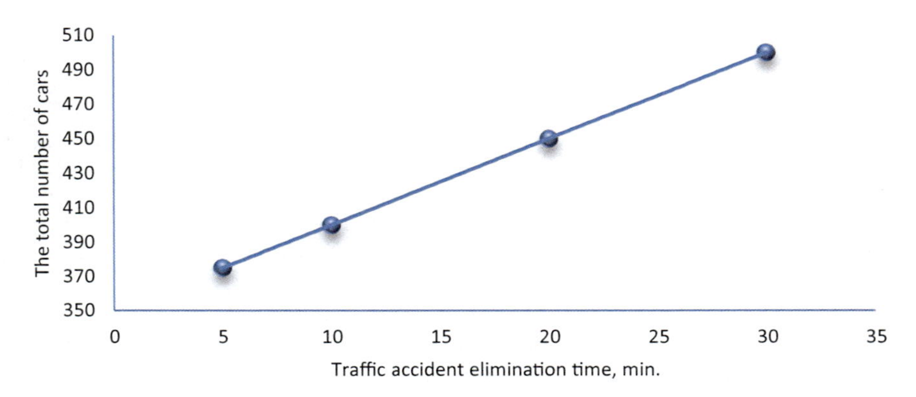

Fig. 8. The total number of cars before traffic accident elimination.

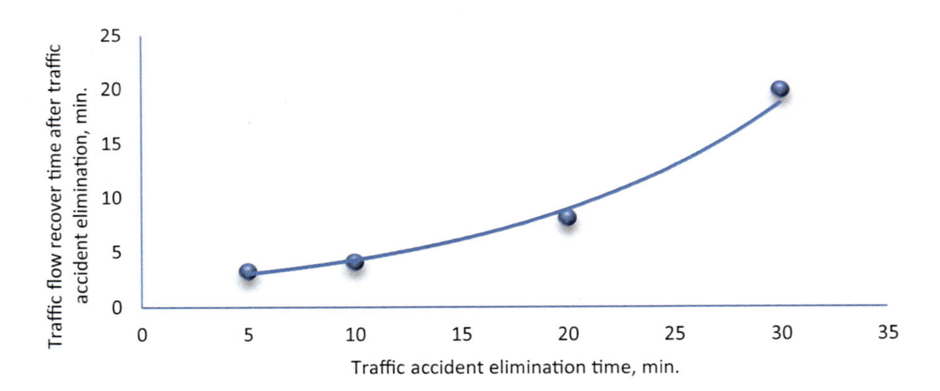

Fig. 9. Traffic flow recovers time after traffic accident elimination.

$$N = 5x + 350 \tag{5}$$

where N – the total number of cars; x – traffic accident elimination time, min.

The longer it takes to eliminate a traffic accident, the greater number of cars is on the modeling road.

The traffic flow after the traffic accident elimination is not restored to its original state immediately, because it becomes unbalanced. And based on the results show that the traffic flow recover time varies according to an exponential function (Fig. 9):

$$y = 2.0562e^{0.0737x} \tag{6}$$

where y – traffic flow recover time after traffic accident elimination, min.; x – traffic accident elimination time, min.

This means that with increasing of the traffic accident elimination time requires more time during which traffic will be restored to its original state.

6 Conclusions

1. Traffic accident elimination time affects the whole traffic flow parameters, causes traffic jams and extends the travel time.
2. After the traffic accident consequences elimination, the traffic flow recovers to the state that was before the accident:
   within 3 min 20 s when the traffic accident elimination time is 5 min;
   within 4 min 10 s when the traffic accident elimination time is 10 min;
   within 8 min 20 s when the traffic accident elimination time is 20 min;
   within 20 min when the traffic accident elimination time is 30 min.
3. Before the traffic accident consequences elimination, the total number of cars increases on road by a linear tendency increasing traffic accident elimination time.
4. In the event of a traffic accident, the formation of traffic congestion can be controlled by changing traffic lights cycles.
5. The discrete traffic flow method can be used to study urban traffic flows with changing conditions in them, e.g., road works, speed limits on individual road sections, the number of lanes and other conditions influence that changes the permeability of the roads under investigation. That results obtained can be used to manage traffic flows by directing traffic flows to another network road.

References

1. Tang, T.Q., Shi, W.F., Huang, H.J., Wu, W.X., Song, Z.: A route-based traffic flow model accounting for interruption factors. Physica A **514**, 767–785 (2019)
2. Gao, J., Dai, L., Gan, X.: Traffic flow and safety analysis. Theor. Appl. Mech. Lett. **8**(5), 304–314 (2018)
3. Ou, H., Tang, T.Q.: Impacts of moving bottlenecks on traffic flow. Physica A **500**(15), 131–138 (2018)
4. Piacentini, G., Goatin, P., Ferrara, A.: Traffic control via moving bottleneck of coordinated vehicles. IFAC-PapersOnLine **51**(9), 13–18 (2018)
5. Tang, T.Q., Li, P., Yang, X.B.: An extended macro model for traffic flow with consideration of multi static bottlenecks. Physica A **392**(17), 3537–3545 (2013)
6. Quek, W.L., Chung, N.N., Chew, L.Y.: An analysis on the traffic processing efficiency of a combination of serial and parallel bottlenecks. Physica A **503**(1), 491–502 (2018)
7. Jiang, R., Jin, C.J., Zhang, H.M., Huang, Y.X., Jia, B.: Experimental and empirical investigations of traffic flow instability. Transp. Res. Part C: Emerg. Technol. **94**, 83–98 (2018)
8. Yuan, S., Zhao, X., An, Y.: Identification and optimization of traffic bottleneck with signal timing. J. Traffic and Transp. Eng. **1**(5), 353–361 (2014)
9. Bogdevičius, M., Junevičius, R.: Investigation of traffic flow dynamic processes using discrete model. J. KONES Powertrain and Transp. **21**(4), 15–19 (2014)
10. Junevičius, R., Bogdevičius, M.: Mathematical modelling of network traffic flow. Transport **24**(4), 333–338 (2009)

Road Safety of Disabled People on the Example of the Campus Equipment

Aleksandr Novikov[1(✉)], Ivan Novikov[2], and Anastasia Shevtsova[2(✉)]

[1] Orel State University named after I.S. Turgenev, St. Komsomolskaya 95, 302026 Orel, Russia
novikovan@ostu.ru

[2] The Federal State Budget Educational Institution of Higher Education «Belgorod State Technological University named after V.G. Shukhov», St. Kostyukova 46, 308012 Belgorod, Russia
{ooows, shevcova-anastasiya}@mail.ru

Abstract. Changing the transportation infrastructure for improving the main characteristics of the transportation flow is the key problem in transportation planning, therefore the main question lies in the ability to plan the change of the main indicators for the long term. In this investigation, an analysis of the city's population has been performed and the most difficult transportation segment has been identified. During its identification, the main characteristics of the transportation flow have been established. For the evaluation of these characteristics until 2025, an analysis of the available methods of establishing changes in their values has been conducted. During the analysis of the above mentioned methods of evaluation of the change in intensity, based on the method of extrapolation, three scenarios of the development of the transportation system have been identified. It has been established that the most favorable method of controlling the transportation flow in the entrance to the city is the long term control of the traffic system. For the first time, with the help of the authors, based on the investigations of foreign scientists and the mathematical analysis of the changes in intensiveness on the main routes of the given road, the method of graphically choosing the required control plan has been put forward. The effectiveness of said organization scheme of the transportation system has been calculated in the software product Transyt-14 software product, with the analysis of changes in the main characteristics of the transportation flow.

Keywords: People with limited mobility · Road safety · Campus · Activities · Economy · Community organizations · The organizational scheme of the transportation system · Forecasting method · Long term · Modelling · Intensiveness of the transportation flow · Values · Characteristics · Graphic method

A. Varhelyi et al. (Eds.): VISZERO 2018, LNITI, pp. 93–103, 2020.
https://doi.org/10.1007/978-3-030-22375-5_11

1 Introduction

People with limited mobility both in the world and in Russia are an important category of citizens who need support and assistance from the state. In the study, in accordance with the existing state programs, measures have been developed to improve the normal life of people with limited mobility on the example of the campus.

In the course of the study, the authors found that people with limited mobility are people who have difficulties in independent movement to obtain various information and services, as well as in orientation in space.

Creating an accessible environment for this category of citizens is an extremely important task, accessible environment is access without barriers to various institutions and objects of urban infrastructure and in particular the territory of the Federal State Budget Educational Institution of Higher Education «Belgorod State Technological University named after V.G. Shukhov» (BSTU). Having organized for them the necessary conditions of accessible environment-it means to take care of them and make the life of these groups more comfortable.

The aim of the study is to improve the safety of movement of people with limited mobility on the example of the campus of BSTU.

The assessment infrastructure of the campus showed that its infrastructure is not adapted to ensure the safety of movement of special groups of the population, which is one of the most important tasks in the implementation of transport planning. In this situation, the increase of traffic safety and implementation of special events with the help of public organization, will allow to reduce social tension, to relieve the burden on a federal portly organizations and foundations, and improve overall road safety.

In earlier studies [1] it was found that in foreign countries similar problems were solved much earlier than in the Russian Federation. The solution of such problems occurs at the legislative level, so the Institute of transport engineers developed a specialized document on the evaluation of new traffic management schemes in 1991 [2]. In his research [3] Theodore offers seven approaches to the evaluation of new objects of movement.

In Russia, as it was said earlier, the problem of using an integrated approach has been approached only now, in this regard, in recent times it has been reflected in a specialized bill [4]. This document reflects the main approaches to the assessment of new road traffic management schemes in the long term.

The main issue in the implementation of new activities and compliance with the requirements prescribed by the legislative document is the forecasting of changes in traffic volume, but based on which it becomes possible to analyze the main characteristics of the transport flow for a long-term period. To implement this approach and develop a graphical method for selecting an effective management regime, the following tasks are solved within the framework of the study:

1. determination of traffic flow parameters at the input sections;
2. quantitative indicators of these characteristics are established using the simulation procedure;

3. the most accurate method for predicting the change in the amount of traffic is determined, which makes it possible to calculate the magnitude of the intensity in the long term, taking into account three possible scenarios for the development of the transport network;
4. a mathematical analysis performed, the functional dependence of the intensity of the transport sweat on time is determined;
5. a graphic method has been developed to select the effective duration of traffic flow control;
6. the simulation procedure was carried out taking into account the proposed method of selecting the management regime and the prospective three scenarios for the development of the changing transport situation at the site.

2 Analysis of the Research Object

The urban settlement "Dubovoe" is a settlement located in the Belgorod region, part of the Belgorod agglomeration, the area of Dubovoy is approximately 13 km^2. According to the Federal Service of State Statistics [5], the population size for the period 2008–2016 is determined.

Over the past 8 years, the population of the settlement has increased by more than 3,000 people, which in turn is 42% of the 2008 figure. Annually, on average, the number of inhabitants of the urban settlement increases by an average of 300 people, and the area of the settlement remains equal to 13 km^2. This phenomenon is explained by the construction of large-scale high-rise residential complexes with affordable price policy, which allows the population to purchase apartments in the urban settlement. In turn, the population growth for the long term will lead to the emergence of transport problems in the urban road network.

The main intersection of the settlement in question is the ring of Shchors st.-Vatutin ave. During the study, the intensity of the node was examined (Table 1).

Table 1. Intensity of traffic flow at the intersection of str. Shchorsa – str. Vatutina in the morning and evening rush hour.

Intersection of str. Shchorsa – str. Vatutina											
Vehicle type								Passenger car equivalent	Aimsun classification		
№ Directions	Passenger cars	Lorries (small)	Lorries (Big)	Share taxi			Trolley-buses				
				Small	Medium	Big			Passengers cars	Lorries	Buses
1	57 (109)	2 (3)	1 (0)	1 (4)	0 (3)	0 (0)	0 (0)	64 (126)	57 (109)	2 (2)	1 (5)
2	76 (210)	2 (13)	0 (0)	5 (1)	3 (12)	1 (5)	0 (0)	95 (268)	76 (210)	2 (8)	0 (15)
3	108 (30)	7 (4)	1 (0)	1 (0)	10 (0)	2 (0)	0 (0)	148 (36)	108 (30)	7 (2)	1 (0)
4	82 (19)	4 (0)	2 (0)	1 (0)	2 (1)	0 (1)	0 (0)	99 (24)	82 (19)	4 (0)	2 (2)
5	59 (112)	0 (3)	5 (2)	0 (0)	0 (0)	0 (0)	0 (0)	72 (122)	59 (112)	2 (4)	5 (0)

(*continued*)

Table 1. (*continued*)

Intersection of str. Shchorsa – str. Vatutina											
Vehicle type								Passenger car equivalent	Aimsun classification		
№ Directions	Passenger cars	Lorries (small)	Lorries (Big)	Share taxi			Trolley-buses				
				Small	Medium	Big			Passengers cars	Lorries	Buses
6	82 (142)	5 (1)	4 (1)	0 (1)	0 (0)	0 (0)	0 (0)	100 (148)	82 (142)	5 (2)	4 (1)
7	187 (100)	12 (7)	1 (3)	0 (0)	0 (0)	0 (0)	0 (0)	208 (118)	187 (100)	12 (7)	1 (0)
8	76 (200)	2 (5)	2 (2)	4 (0)	2 (4)	1 (1)	0 (0)	97 (223)	76 (200)	2 (5)	2 (4)
9	53 (80)	4 (3)	1 (0)	0 (4)	0 (2)	0 (0)	0 (0)	62 (95)	53 (80)	4 (2)	1 (4)
10	154 (149)	9 (4)	4 (2)	0 (0)	0 (0)	0 (0)	0 (0)	178 (160)	154 (149)	9 (4)	4 (0)
11	150 (73)	7 (10)	2 (6)	0 (0)	0 (0)	0 (0)	0 (0)	166 (103)	150 (73)	7 (12)	2 (0)
12	42 (69)	1 (3)	2 (0)	0 (4)	0 (2)	0 (0)	0 (0)	49 (95)	42 (80)	1 (2)	2 (4)

The parameters of the main flow are determined during the simulation of the investigated node (Table 2).

Table 2. Flow parameters the intersection of str. Shchors – str. Vatutina.

Main characteristics of the traffic flow	Morning	Evening
Delay, s/km.	122.91	280.79
Speed, km/h.	25	17
Cars in the cycle, vehicles/h.	3.982	4.202

During the simulation, it is established that the values of the characteristics of the flow approaching 1, leading to the emergence of "bottleneck". In total of researches, it is established that it is necessary to carry out organizational actions.

3 Definition of the Approach to the Study

To implement the predictive approach, earlier [1] the analysis of scientific sources devoted to this problem [6].

Thus, in his research Hashem [7] uses a linear forecast of the increase in the intensity of traffic flow, on the contrary Dixon [8] uses a predictive method only in the development of complex traffic and claims the need for forecasting. Research carried out at the University of Indonesia [9] suggests the need for the use of geographic information systems, but there is a complexity of their application in practice.

Russian scientist, Dinges [10] explores two methods for predicting:

1. Method «moving average» . Forecasting the traffic intensity using simple moving averages is advisable to use when forecasting on secondary roads, on certain sections of roads or in places with low population density and low mobility of vehicles [11].

2. Weighted Moving Average Method. The essence of this method is that we assign to each value in a year a certain specific weight (expressed in fractions of a unit), thereby assessing the significance of this year on the value of the predicted parameter, in connection with certain events of those years (be it a cataclysm or a sharp population growth, In connection with the construction of a new microdistrict).

According to the analysis of the scientific literature, the method of extrapolation is best used in the conditions of the existing network of highways, therefore, this method is determined to calculate the prospective traffic intensity on the transport highway in question and the subsequent calculation of necessary control plans.

4 Calculation of Predicted Value of Traffic Intensity

An analysis of the growth in the intensity of vehicles in the urban settlement using the extrapolation method, in particular the weighted moving average method, made it possible to calculate the main scenarios for the development of car growth up to 2025, taking into account the possible increase in the number of cars in three possible scenarios:

1. Minimal – the forecasted percentage of growth will be 129.63%, taking into account the fact that the increase in the number of cars will not be affected by urban development. In the urban settlement, construction of new residential complexes and objects of attraction of the population will not be planned;
2. Average – the forecasted percentage of growth will be 161.96%, taking into account the fact that the increase in the number of cars will be influenced by urban development. In the urban settlement, construction of new residential complexes and objects of attraction of the population by 50% of the total area under construction will be made;
3. Maximum – the projected percentage of growth will be 194.3%, taking into account the fact that the increase in the number of cars will be influenced by urban development. In the urban settlement will be built new residential complexes and objects of attraction of the population by 100% of the total area ready for building.

Due to the fact that in the calculation of the possible increase in intensity at the study intersection were used the data obtained as a result of multidimensional research (specific values), the result of growth under three possible scenarios were obtained in a certain (precise) values (129.63%, 161.96%, 194.30%).

Based on the analysis of three possible scenarios for the development of the urban transport network, an imitation procedure was performed, taking into account the possible increase in intensity (Table 3).

Table 3. Flow parameters in the course of service-simulating test of existing scheme of traffic management at the intersection of str. Shchorsa – str. Vatutina.

The main characteristics of traffic flow	Morning				Evening			
	Actual traffic intensity	129.63%	161.96%	194.30%	Actual traffic intensity	129.63%	161.96%	194.30%
Delay time, s/km	122.91	250.88	269.19	263.9	280.79	422.49	404.55	411.1
The number of vehicles that passed the intersection, vehicles/h	3.982	4.161	4.388	4.724	4.202	4.142	4.309	4.255
Traffic flow average speed, km/h	25.99	19.23	17.94	17.1	17.72	13.14	13.59	13.23

It is established that the main indicators, especially during the peak periods, will be significantly reduced, therefore, in order to assess the functioning of the node under consideration, the necessary measure is the definition of a new scheme of traffic organization. The only activity in which the distribution of traffic flows in time and their separation, in order to avoid occurrence of harsh situations, is the introduction of traffic light regulation [12, 13]. The main issue is to determine the optimum duration of control, in this regard, the author's team performed a mathematical analysis of the dependence of traffic flow intensity on the node under consideration and the method of choice of the control plan - graphic is offered.

5 Method of Selecting the Required Management Plan

To develop the method, the dependence of the most intensive approach on the total indicator is determined:

$$N_{dir} = f(N_{int}, \tau) \tag{1}$$

where N_{dir} – intensity of vehicles in the direction; N_{int} – total traffic flow intensity; τ – time value.

Dependences of intensity indicators of the movement in each of the directions:

$$f(\tau; n) = a + \frac{b}{\tau} + cn + \frac{d}{\tau^2} + en^2 + \frac{fn}{\tau} + \frac{g}{\tau^3} \tag{2}$$

where τ – time value; n – intensity of car flow 1; a = 804.97; b = −1131.38; c = 0.24; d = 104.42; e = −0.0008; f = 1,72; g = 16.01; i = −0.0003; j = −0.18; R^2 = 0.97.

This approach is used to determine the parameters R^2 and R^2adj using reduction factors [14]. Their values are shown in Table 4.

In the course of the mathematical description, the influence of the difference in intensity in directions on the mode of motion is determined, which is confirmed by the functional dependence [15–17]. To improve the efficiency of the organization of movement on the object under study, the authors propose to use traffic light control using a graphical approach to the definition of the control plan. As a result of the scientific review [18–22], a graphical method of selecting the most loaded lane for traffic was developed.

Table 4. Values of mathematical review.

Coefficient	Direction 1	Direction 2	Direction 3	Direction 4
a	804.97	−139.87	−2.91E+06	−4704.12
b	−1131.38	2267.46	57227.2	36515.14
c	0.24	9837.31	−196565.25	−41283.03
d	104.42	16952.8	327694.99	−61624.8
e	−0.0008	12160.45	−263986.47	1.28E+09
f	1.72	3032.22	81851.3	−1.01E+07
g	16.01	1.45	2.18E+06	37881.22
R^2	0.97	0.97	0.97	0.97
R^2_{adj}	0.865	0.865	0.865	0.865
AIC	2.781	2.781	2.781	2.781
Schwartz	2.441	2.441	2.441	2.441

This way can be developed on the basis of analyses of differences of intensity on the analyzed knot, within week, month, year and identification of zones for use of identical operating modes of a traffic light object. Expedient use of operating modes of a traffic light object will allow to increase capacity, to reduce inadvertent stops of vehicles on crossing that will lead to reduction of exhaust gases and economy of fuel.

The use of data filled with one approach it is possible to obtain the result for the rest, like the classical approach of determining the cycle:

$$T_c = T_{c(max)} \tag{3}$$

where T_c – time cycle (taking into account all directions), sec; $T_{c\ (max)}$ – time cycle (accounting for the loaded phase direction), sec.

A certain regularity allows to optimize the calculation of the optimal duration depending on the most loaded direction, which greatly simplifies the calculation.

It is proposed to present a 3-phase control system in the form of a graphical approach (see Fig. 1).

Mathematical description:

$$\begin{Bmatrix} X \\ Y \\ Z \end{Bmatrix} = (X, Y, Z) \overset{max}{\rightarrow} mode[n] \tag{4}$$

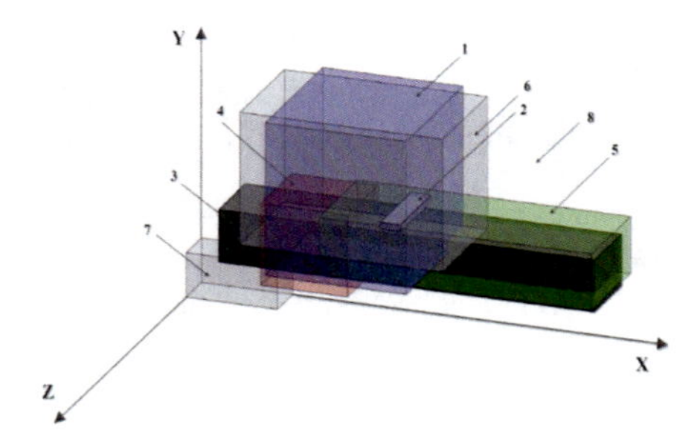

Fig. 1. Selection method – graphic on the object: 1 – mode 1; 2 – mode 2; 3 – mode 3; 4 – mode 4; 5 – mode 5; 6 – mode 6; 7 – do not use traffic lights; 8 – further calculation.

where X – loaded direction in the first phase, un/h; Y – loaded direction in the second phase second phase, un/h; Z – loaded direction in the second phase third phase, un/h; n – management plan (1…6) (Table 5).

Table 5. Identified through the mathematical analysis of the maximum and minimum intensity values for each control mode.

Mode 1	*X*	*Y*	*Z*
MAX the value of the intensity in the direction	1416	1164	750
MIN the value of the intensity in the direction	726	480	341
mode 2	X	Y	Z
MAX the value of the intensity in the direction	1265	666	398
MIN the value of the intensity in the direction	1167	635	365
mode 3	X	Y	Z
MAX the value of the intensity in the direction	2133	537	348
MIN the value of the intensity in the direction	287	303	187
mode 4	X	Y	Z
MAX the value of the intensity in the direction	932	602	402
MIN the value of the intensity in the direction	507	236	164
mode 5	X	Y	Z
MAX the value of the intensity in the direction	2197	915	504
MIN the value of the intensity in the direction	858	387	240
mode 6	X	Y	Z
MAX the value of the intensity in the direction	1560	1158	579
MIN the value of the intensity in the direction	564	564	411

Using a graphic method calculation of necessary duration of the mode of regulation is executed and the analysis with the existing mode (without use of traffic light management) according to the main characteristics of the traffic flow received during modeling is made: stops and delays; turns and jams; fuel consumption.

The indicator of efficiency has on average improved on all listed indicators for 12%. Network results of consumption of fuel at the existing and received mode are presented in the Table 6.

Table 6. Network of fuel consumption.

Time period		Fuel consumption as a result of the delay (liters per hour)	Fuel consumption as a result of stops (liters per hour)	Total fuel consumption (liters per hour)
08:00-09:00 (before (now))	260.99	134.29	130.34	525.62
08:00-09:00 (after)	210.99	121.02	109.15	495.75

When comparing the obtained parameters, it was found that in the course of using the proposed method of choosing the optimal control plan of the object of study, taking into account the forecast scenario of development, the transport situation will improve as a result of changes in the value of the delay of cars (Table 7).

Table 7. Parameters of the flow in the course of service-simulating test of existing scheme of traffic management at the intersection of str. Shchorsa – str. Vatutina taking into account intensity growth forecast.

The main characteristics of traffic flow	Morning				Evening			
	Actual traffic intensity	129.63%	161.96%	194.30%	Actual traffic intensity	129.63%	161.96%	194.30%
Delay time, s/km	87.94	96.57	182.21	291.92	57.14	144.25	254.64	314.57
Number of cars driven, vehicles/h	3979	5255	6399	6421	4464	5893	6660	6647
Flow rate, km/h	36.66	33.45	27.66	21.85	35.54	27.06	19.94	17.22

6 Conclusions

Studies of the object of the transport network have shown that the high growth of transport characteristic of many cities creates problems on the transport network, especially at the entrance areas. So the study of one intersection showed that the load on it is significant and is expressed in the reduced characteristics of the traffic flow, namely the speed (< 25 km/h) and the delay time (280 s/km (max!)). The authors in the course of research offers a scientific approach to improving the transport situation, concluding in the development of a predictive approach to the analysis of the land transport network and the development of a graphical method of determining the optimal management plan. As a result of testing the proposed approach at the research site, it is possible to achieve positive results, namely, an increase in the speed of movement (35 km/h) and a decrease in the delay time (87 s/km).

The relevance of the chosen study is confirmed by the high load on the road network, which is currently typical for many European countries and the Russian Federation. The introduction of intelligent transport systems makes it possible to manage traffic flows and thus reduce the load, but their use without specialized scientific literature is impossible, so it is necessary to pay special attention to new scientific approaches to the use of such systems and the development of scientific applications for their implementation. The author's team proposed a new approach to road network management using a mathematical apparatus for accounting for the development of basic characteristics in different periods (short, medium and long-term), automatic accounting of flow parameters and the use of intelligent transport systems allows to obtain an effective movement of vehicles using forced control, thus it is possible to minimize delays and reduce the risk of an emergency.

References

1. Novikov, A., Katunin, A., Novikov, I., Kravhenko, A., Shevtsova, A.: Development of a graphical method for choosing the optimal mode of traffic light. J. Phys. Conf. Ser. **1015**(3) (2018)
2. Brian, S.: Traffic Access and Impact Studies for Site Development, a Recommended Practice, Prepared by the Transportation Planners Council Task Force on Traffic Access. Institute of Transportation Engineers, Washington, DC (1991)
3. Teodoro, R., Regin, J., Val, R.: Traffic impact assessment for sustainable traffic management and transportation planning in urban areas. In: Proceedings of the Eastern Asia Society for Transportation Studies, vol. 5, pp. 2342–2351 (2005)
4. Decree of the Ministry of Transport of the Russian Federation No. 43 of March 17, 2015. On Approval of the Rules for the Preparation of Road Traffic Projects and Traffic Schedules. Registered with the Ministry of Justice of the Russian Federation on 17 June 2015. Registration No. 37685 (2015)
5. GKS Homepage. http://www.gks.ru/. Accessed 21 May 2017
6. Morichi, S.: Long-term strategy for transport system in Asian mega-cities. J. East. Asia Soc. Transp. Stud. **6**, 1–22 (2005)
7. Al-Masaeid, H.R., Al-Omoush, N.J.: Traffic volume forecasting for rural roads in Jordan. Jordan J. Civ. Eng. **159**(3147), 1–13 (2014)
8. Dixon, M.: The effect of errors in the annual average traffic forecasting: study of highways in rural Idaho. Report budget no. KIK253, 4–12, 1–60 (2004)
9. Chaleb, O., Sediyono, E.: Forecasting the case of traffic accidents through the geographic information system (GIS) application method with double exponential smoothing and analytical hierarchy process (AHP) in the city of Jayapurapapua. J. Theor. Appl. Inf. Technol. **83**(3) (2016)
10. Dinges, E.: Methods for Planning and Evaluating the Effectiveness of Measures to Improve road Safety, p. 140. Moscow Automobile and Road Technical University (MADI), Moscow (2016)
11. Vlasov, V., Novikov, A., Novikov, I., Shevtsova, A.: Definition of perspective scheme of organization of traffic using methods of forecasting and modeling. In: IOP Conference Series: Materials Science and Engineering (2018)

12. Kerner, B.: Introduction to Modern Traffic Flow Theory and Control: The Long Road to Three-Phase Traffic Theory. Springer, Berlin, New York (2009)
13. Roess, R., Prassas, E., McShane, W.R.: Traffic Engineering, 4th edn, p. 744. Prentice Hall, Upper Saddle River (2010)
14. Makridakis, S., Hibon, M.: The M-3 competition: results, conclusions and implications. Int. J. Forecast. **16**, 451–476 (2000)
15. Rich, J., Bröcker, J., Hansen, C.O., Korchenewych, A., Nielsen, O.A., Vuk, G.: Report on Scenario, Traffic Forecast and Analysis of Traffic on the TEN-T, taking into Consideration the External Dimension of the Union – TRANS-TOOLS version 2; Model and Data Improvements, Funded by DG TREN, Copenhagen, Denmark, p. 133 (2009)
16. Xianfeng, Y., Chang, G., Rahwanji, S., Lu, Y.: Development of planning-stage models for analyzing continuous flow intersections. J. Transp. Eng. **139**(11), 1124–1132 (2013)
17. Wong, C., Wong, S.: Lane-based optimization of signal timings for isolated junctions. Transp. Res. Part B Methodological **37**(1), 63–84 (2003)
18. Schutter, B., Bart, L.: Optimal Traffic Light Control for a Single Intersection. In: Proceedings of the American Control Conference San Diego, California, pp. 2195–2199, June1991
19. Yanfeng, G., Christos, G.: Cassandras multi-intersection traffic light control with blocking. Discrete Event Dyn. Syst. **25**, 7–30 (2015)
20. Baskar, L., Schutter, B., Hellendoorn, H.: Traffic management for automated highway systems using model-based predictive control. IEEE Trans. Intell. Transp. Syst. **3**(2), 838–847 (2012)
21. Broucke, M., Varaiya, A.: A theory of traffic flow in automated highway systems. Transp. Res. C **4**(4), 181–210 (1996)
22. Prashanth, L., Bhatnagar, S.: Reinforcement learning with function approximation for traffic signal control. IEEE Trans. Intell. Transp. Syst. **12**(2), 412–421 (2011)

Road Traffic Management During Special Events

Dovydas Skrodenis(✉)

Road Research Institute, Vilnius Gediminas Technical University,
Saulėtekio al. 11, 10223 Vilnius, Lithuania
dovydas.skrodenis@vgtu.lt

Abstract. Planned special events (PSE) every year attract many people to one place. After all, increased traffic volume cause congestions and has a negative impact on the economy by decreasing productivity and decreasing the quality of people's lives. Nevertheless, environmental pollution increases with increasing traffic. The effects of traffic congestion on traffic safety, however are less obvious. Fatal accident risks are less common in traffic jams but the probability of severe rear end crashes is increased. Investigation of real case study in Lithuania has been made and results are discussed in this paper.

Keywords: Traffic volume · Alternate routes · Planned special events

1 Introduction

Mobility has been increasing significantly in the last decades and will continue to increase [3]. In Lithuania, car use increased by 12% between 2014 and 2017 [1].

Road stretches which have insufficient capacity for increased traffic cause traffic congestion and delays. Traffic congestion has a negative impact on the economy by decreasing productivity and decreasing the quality of people's lives [5]. Besides that, environmental pollution increases with increasing traffic (fuel consumption and noise towards to increased emissions).

The effects of traffic congestion on traffic safety, however, are less obvious. Rietveld and Shefer [2] suggest that congestion might have a positive effect on safety by decreasing the number of fatalities as speeds decrease [5]. Although this statement seems logical, when looking at the traffic conditions in more detail the effects of congestion on safety are less apparent [5]. As traffic flow increases and density approaches its critical value, traffic flow is said to be unstable [3]. Any minor disruption may lead to crashes under these conditions. The formation of traffic jam can cause rear-end crashes at the end of the queue because of large differences in speed [5]. Oh et al. [4] after the study suggests that the increased traffic speed variation leads to more crashes. Furthermore, motorways give evidence of structural congestion cause road users to seek alternative routes which are not suitable for high traffic volume and with a higher accident risk. It is important to gain a clear understanding of how traffic flow processes affect safety, in order to better understand developments in traffic safety and identify possibilities for improvement [3].

A. Varhelyi et al. (Eds.): VISZERO 2018, LNITI, pp. 104–109, 2020.
https://doi.org/10.1007/978-3-030-22375-5_12

The largest traffic flows in the country arise in motorways or highways. These roads are designed for fast and safe transport communication, but sometimes due to certain circumstances, these roads become hard to pass as the result of the traffic congestion. In most cases, road users tend to avoid congestion and opt for alternative roads that are not adapted to high traffic volumes, and the traffic safety measures applied there are not sufficient to ensure a sudden increase in vehicle intensity. There are two types of congestion: structural or incidental [7]. Structural congestion occurs when traffic demand is higher than capacity, while incidental congestion is the result of occasional conditions such as a crash, bad weather or road works which alter the traffic flow [5, 7]. Incidental congestion cases cannot be predicted and occur naturally, but structural congestion may occur due to improper road maintenance or inappropriate traffic management.

2 Traffic Management Issues During PSE

Each country has its own traditions and national events that take place every year. However, occasional cases of large-scale events are often made irregularly and their organization causes some challenges in the communication scheme. For illustration of the scale of what problems municipalities faces, approximately 1,000 planned special events happens in Los Angeles Country each year and these challenges are not limited to only the largest metropolitan areas [6]. More than five million people attend PSE in Milwaukee, Wisconsin each year. Each year, over 400 planned special events are the main cause of the traffic growth along the I-94 freeway corridor in downtown Milwaukee [6]. PSE have impact to travel safety, mobility, and travel time reliability across all surface transportation modes and roadway facilities [6]. Advanced planning operations of stakeholders and their coordination is needed for multi-agency transportation management plan development [6].

PSE increase traffic demand and reduce roadway capacity for this reason travelers faces unexpected delays and congestions. These delays are unlike congestion caused by routine traffic during daily peak travel periods [5]. The frequency and severity of traffic disruptions due to PSE is increasing significantly in metropolitan area and suburbs. It is necessary to predict the potential hazards and the amount of traffic flow when organizing road closures (e.g., road works) in order to avoid as much traffic as possible. In most cases, normal traffic flow is evaluated and certain criteria are not taken into account.

PSE affect all road users and service providers who are using adjoining roadway network. Law enforcement, medical or fire responders, local or regional transit experience time delays and unexpected congestion phenomenon [6].

However, planned special events is known object of attraction for local or public agencies with defined time, duration, location and other relevant information. All this advanced data can be used to minimize the impact of PSE for various types of road users and service providers. Intelligent Transportation Systems (ITS) often are chosen to apply traffic management techniques to minimize the rate of unexpectedness [5].

3 Case Studies

Several national events are held every year on the seaside, attracting a lot of people from all over the Lithuania. One event called "Palanga Stint" held at the end of February, another event "Sea Festival" – at the end of July. Traffic volume data is taken from road-mounted sensors. To distinguish how much traffic volume changes, comparison of data between typical weekend and celebration weekend has been made.

Detailed traffic volume analysis revealed that the number of vehicles towards Klaipėda city increased by 9.69% in 2016 and 13.38% in 2015. Looking at outwards Klaipėda city traffic volume numbers the increase is visible – 13.38% in 2016 and 5.35% in 2015.

These numbers prove the statement that mobility is constantly increasing. Road No. A1 is highway constructed for the main traffic flow towards and outwards Klaipėda city from the capital and other big cities. However, traffic volume data from surrounding roads revealed, that road users tend to choose alternative roads to reach Klaipėda especially during national events. Figures 1, 2 represents detail traffic volume statistics during "Sea Festival" and regular weekend.

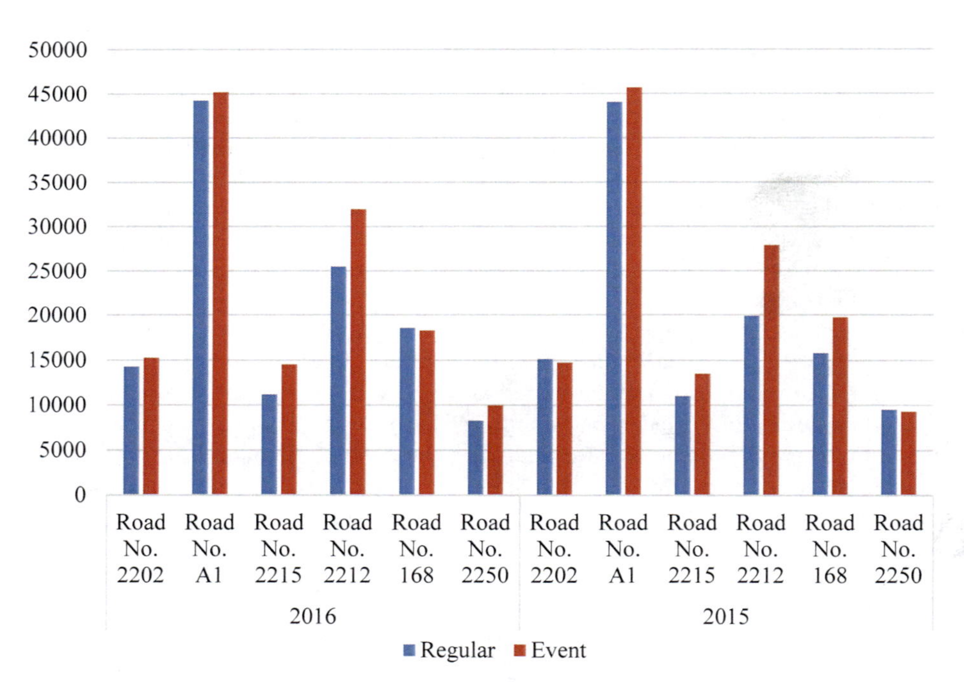

Fig. 1. Comparison of traffic volume towards Klaipėda city during "Sea Festival" and regular weekend.

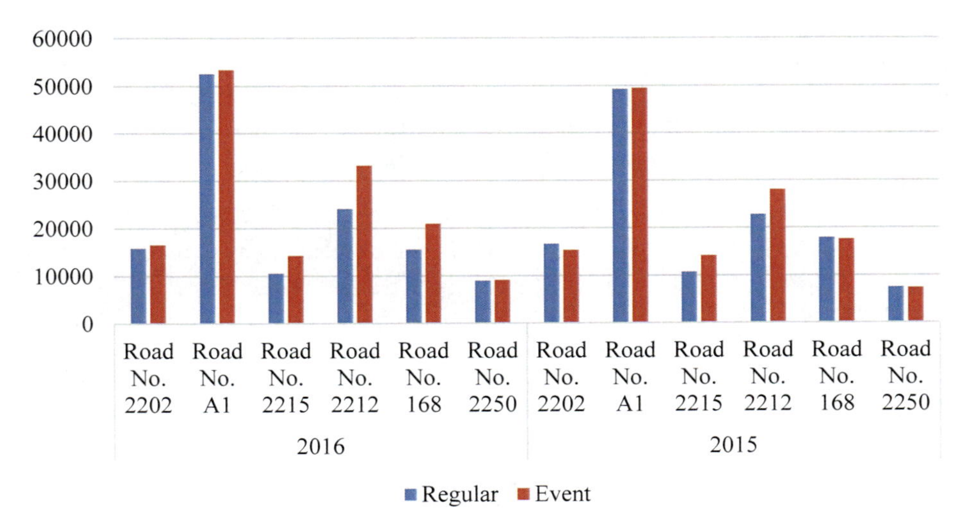

Fig. 2. Comparison of traffic volume outwards Klaipėda city during "Sea Festival" and regular weekend.

Regarding to the statistics in Table 1, regional roads are often chosen as alternative route towards and outwards Klaipėda city. Regional roads are constructed for low traffic volumes and not suitable to serve as the main road. However, planned special events distorts traffic management plans as roads users seeks for alternate roads to reach their destination faster. Numbers in Table 1 represents the difference of traffic volumes in different roads between "Sea Festival" and regular weekends. More than 135,200 vehicles entered the Klaipėda city in 2016 and 131,000 in 2015 during "Sea Festival" event. Highway A1 is the main road to reach Klaipėda and generates one third of all traffic volume to the city, however looking at the traffic volume statistics of 2 years, more and more roads users choose regional roads towards Klaipėda. 50.01% of roads users entered Klaipėda using regional roads in 2015 (event weekend) while in 2016 number has increased to 53.05%.

Table 1. Difference between traffic volume towards and outwards Klaipėda during "Sea Festival" and regular weekend.

Road No.	2202	A1	2215	2212	168	2250
Towards Klaipėda						
2016	6.35%	2.08%	23.02%	20.14%	−1.63%	17.09%
2015	−2.75%	3.65%	18.45%	28.62%	19.92%	−2.47%
Outwards Klaipėda						
2016	4.24%	1.47%	25.69%	27.28%	25.76%	1.55%
2015	−8.72%	0.37%	24.03%	18.71%	−1.70%	−1.73%

Regional roads No. 2215, No. 2212 and No. 2250 are the most popular ways to reach Klaipėda during “Sea Festival”. In 2016, the intensity of these roads comparing to regular weekend increased by 21.31% (Road No. 2215), 19.76% (Road No. 2212) and 17.29% (Road No. 2250). Despite the fact, that these roads go towards Klaipėda, they also connect small villages and the increased traffic volume could result a higher accident possibility.

Another big event which attracts many people to one place is “Palanga Stint”. This event attracts from 29% to 47% more road users while comparing to regular weekend. This event happens in the middle of winter every year and increased traffic flow determines that road maintenance works should be done correctly to reduce the possibility of traffic accidents (Fig. 3).

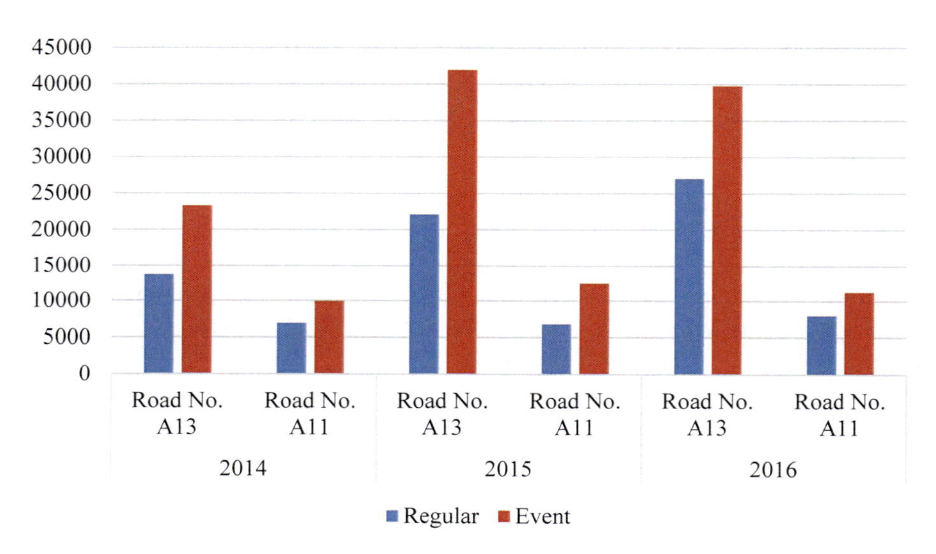

Fig. 3. Comparison of traffic volume towards Palanga city during “Palanga Stint” and regular weekend.

4 Conclusions

Unlike traffic incidents or natural disasters, planned special events attract road users to known place and at known time. Public agencies should expect increased traffic volume and prepare traffic flow management plans during the event.

“Sea Festival” attracts 9.69% (in 2016) and 13.38% (in 2015) more road users to Klaipėda city if comparing to regular weekend. 33.43% (in 2016) and 34.90% (in 2015) of road users choose highway to reach the city, while 53.05% (in 2016) and 50.01% (in 2015) choose regional roads.

During the “Sea Festival” event, regional roads are often chosen as alternative way to reach Klaipėda. In 2016, the intensity of these roads comparing to regular weekend increased by 21.31% (Road No. 2215), 19.76% (road No. 2212) and 17.29% (road No. 2250).

"Palanga Stint" attracts from 29% to 47% more road users while comparing to regular weekend. The number depends on the road and what cities it connects.

References

1. Lithuanian official statistics portal. https://www.stat.gov.lt/en. Accessed 12 May 2018 (2018)
2. Rietveld, P., Shefer, D.: Congestion and safety on highways: towards an analytical model. In: Proceedings of the Third International Conference on Safety and the Environment in the 21st Century (1994)
3. Marchesini, P., Weijermars, W.: The relationship between road safety and congestion on motorways. SWOV Inst. Road Saf. Res., 28 (2010)
4. Oh, C., et al.: Real-time estimation of freeway accident likelihood. In: Proceedings of the 80th TRB Annual Meeting 2001. Transportation Research Board TRB, Washington, DC (2001)
5. Duivenvoorden, K.: The relationship between traffic volume and road safety on the secondary road network; a literature review. D-2010-2. SWOV Institute for Road Safety Research, Leidschendam (2010)
6. U.S. Department of Transportation, Federal Highway Administration. https://ops.fhwa.dot.gov/aboutus/one_pagers/planned_events.htm. Accessed 12 May 2018
7. Haragos, I.M., Holban, S., Cernazanu-Glavan, C.: Determination of quality factor used in road traffic. An experimental study. In: IEEE 12th International Symposium on Applied Machine Intelligence and Informatics (SAMI) (2014)

Implementation of the Road Traffic Safety Concept in Belarus

Denis Kapsky[1(✉)], Sergey Bogdanovich[2], and Aleksandra Volynets[2]

[1] Belarusian National Technical University (BNTU), Nezavisimosti Avenue - 65, 220013 Minsk, Republic of Belarus
d.kapsky@gmail.com

[2] Belarusian Road Research Institute "BeldorNII", 4th Zagorodny Lane - 60, 220073 Minsk, Republic of Belarus
bsw001@gmail.com, briefly.g@gmail.com

Abstract. For more than 10 years, there has been a decrease in the number of road traffic accidents in Belarus, as well as in the number of fatalities and injuries in them. Compared to 1998, in 2017 the number of fatalities decreased by a factor of 3.1. The positive dynamics was achieved largely due to the adoption of the Road Traffic Safety Concept in 2006. The Concept envisaged the implementation of a set of engineering measures aimed at improving the road infrastructure.

The absolute values associated with the accident rate are of little use when it is necessary to compare several countries. For this purpose, the total number of fatalities in relation to the population size, to the car fleet size, and some other indicators have been applied internationally. Using these indicators, the situation in Belarus does not look optimal. The number of fatalities per 1 million cars in Belarus is higher than in most of the neighboring countries of the EU and significantly higher than in the leading countries in the sphere of road traffic safety, including those which have adopted the Vision Zero policy. At the same time, starting from 2011, the accident rate reduction in Belarus has slowed down. It can be assumed that the accident rate asymptotically approaches the limit in accordance with Smeed's Law, according to which the fatality rate in road traffic accidents per car fleet unit decreases as motorization increases.

Remaining within the traditional for Belarus engineering activity, which mainly amounts to road maintenance, it is unlikely that road traffic safety will be significantly improved. Based on the analysis of international road traffic safety, as well as taking into account the local specific features, the article discusses measures, the implementation of which will allow to achieve a qualitatively new level of road traffic safety.

Keywords: Road traffic safety · Road traffic accident · Accident rate · Engineering measures · Road infrastructure

1 Introduction

The road is used for moving freight and passengers. However, traffic includes a number of threats such as emergency, economic, environmental, and social threats.

A. Varhelyi et al. (Eds.): VISZERO 2018, LNITI, pp. 110–119, 2020.
https://doi.org/10.1007/978-3-030-22375-5_13

For road users, accidents are the most sensitive threat, since they are directly related to the life, health, and well-being of people. The need for safety is a basic human need, and people have the right to expect safety from the road as well. However, more than 1.3 million people die and about 50 million are injured on the roads around the world every year [1–3].

After Belarus gained independence, the number of cars in the country began increasing rapidly. This process took place alongside the decline in population size. At the same time, the number of accidents on the roads and the number of fatalities in them remained high in the late 1990s and in the early 2000s.

A quite large number of fatalities in road accidents, as well as the lack of positive dynamics of decline of this indicator against the background of the growing number of cars in the early 2000s, contributed to the adoption by the government of the Road Traffic Safety Concept in the Republic of Belarus in 2006.

2 Measures of the Road Traffic Safety Concept in the Republic of Belarus

The Concept envisaged the improvement of state policy in the field of road traffic and its safety. A large set of measures was envisaged in relation to the road infrastructure and traffic management.

The following was envisaged in relation to the road infrastructure:

- improving the level of design, construction, repair and maintenance of the road infrastructure;
- planned construction of road barriers, artificial road lighting, particularly at dangerous pedestrian crossings, construction of parking lots for vehicles, road junctions, underground and overground pedestrian crossings;
- re-equipment of intersections by creating roundabouts, additional acceleration and deceleration lanes;
- work on the elimination of potentially dangerous sections of roads, and increasing the awareness of drivers and pedestrians about dangerous traffic conditions;
- creating the necessary conditions for cyclists to participate in road traffic, building and equipping bicycle lanes and bicycle parking lots;
- improvement of the video surveillance system that records the traffic situation on main streets in cities, as well as on the republican roads.

The following measures were planned in relation to traffic management:

- timely identification and elimination of the causes and conditions that contribute to the violation of the requirements of technical standards in the field of road traffic and its safety;
- optimization of speed limits for vehicles;
- ensuring the necessary conditions for the optimal movement of pedestrians, especially people with disabilities and children;
- improving methods for calculating losses in road traffic;

- system analysis and improvement of standards defining the requirements for the regulation of road traffic;
- improving the conditions and procedure of participation of heavy and large vehicles in road traffic.

In order to develop the requirements of the Concept, the efforts of the national road administration were aimed at the implementation of the following engineering measures, first of all on primary roads:

- design and construction of by-passes around settlements. Over the past 10 years, 13 such projects have been implemented;
- construction of transport junctions at different levels, minimization of the number of off-ramps and intersections, construction of rotary intersections;
- separation of traffic and pedestrian flows by means of constructing overground and underground pedestrian crossings;
- installation of road barriers, traffic signs and traffic lights, horizontal road marking,
- regulation of speed limits and traffic flow composition;
- protection of roads from snow;
- installation of artificial lighting on dangerous sections and in settlements;
- using anti-dazzle devices, if necessary, to eliminate headlamp glare from the oncoming traffic;
- preventing wild animals from walking onto the roadway.

Since the adoption of the Concept, work has been deliberately carried out to eliminate potentially hazardous road sections, while at the same time measures have been taken to increase the awareness of drivers and pedestrians about dangerous traffic conditions.

There were 749 high accident concentration sections in the country in 2005, and 265 such sections in 2015, 65 of which were left from 2005, and 200 of which appeared from 2005 to 2015. Accordingly, 684 sections were eliminated from 2005 to 2015.

Today it can be said that the implementation of the measures of the Concept has yielded positive results. Between January and December 2017, 3,418 road accidents involving injured people were reported in Belarus, in which 589 people died and 3,620 people were injured. Compared to 1998, when 1,832 people died in road accidents, the number of fatalities decreased by 3.1 times. Starting from 2003, with a constant increase in the number of vehicles, there has been a general tendency of a decrease in the total number of road accidents, as well as in the number of fatalities (Fig. 1).

The chart in Fig. 1 allows concluding that after 2006, there has been a general tendency of a decrease in the number of fatalities in the Republic of Belarus. The data make a rather favorable impression at first glance. However, when comparing them with accident statistics of other countries, it is clear that the situation is quite alarming.

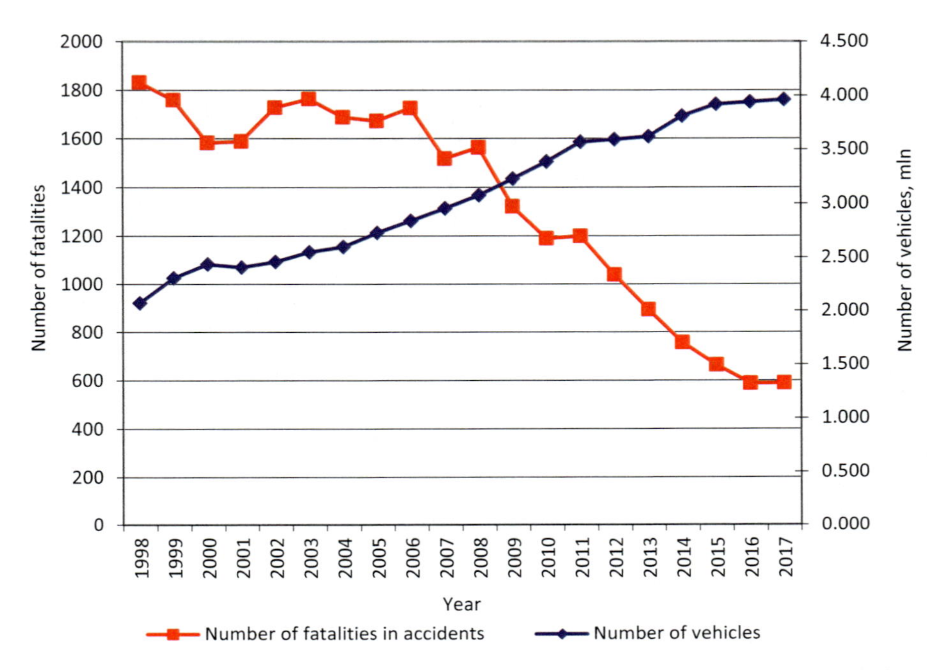

Fig. 1. Change in the number of vehicles and fatalities in road accidents in the Republic of Belarus.

3 Situation in Belarus and Global Accident Rate Trends

The absolute values associated with the accident rate are of little use when it is necessary to compare several countries. For this purpose, such indicators as the total number of fatalities in relation to the population size (social risks), to the car fleet size (transport risks), and some others have been applied internationally. Using these indicators, the situation in the Republic of Belarus does not look optimal.

Figure 2 shows the number of fatalities per 1 million vehicles in the Republic of Belarus, and Fig. 3 shows the same data for countries with traditionally high levels of traffic safety development.

Analysis of the charts in Figs. 2 and 3 allows making several conclusions. The number of fatalities per 1 million vehicles in the Republic of Belarus is higher than in most of the neighboring EU countries and significantly higher than in the leading countries in the field of traffic safety (Sweden, Netherlands, UK, Australia, etc.). At the same time, the considered indicator of the leading countries has changed very little over the course of 20 years, as a matter of fact remaining within a fairly narrow corridor. The explanation of this fact was first suggested by Reuben Smeed as early as 1949. He connected the accident statistics with the level of motorization in the country and showed that the fatality rate in road accidents per car fleet unit decreases as motorization increases [4, 5]. The dependencies proposed by him, called Smeed's Law, were subsequently elaborated and repeatedly tested in many countries. At the same time, the

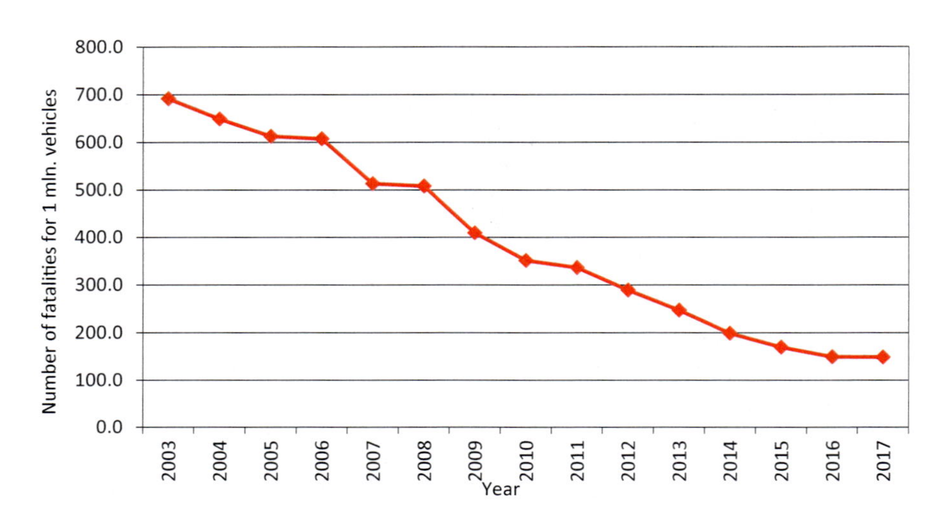

Fig. 2. Change in the number of fatalities in road accidents per 1 million vehicles.

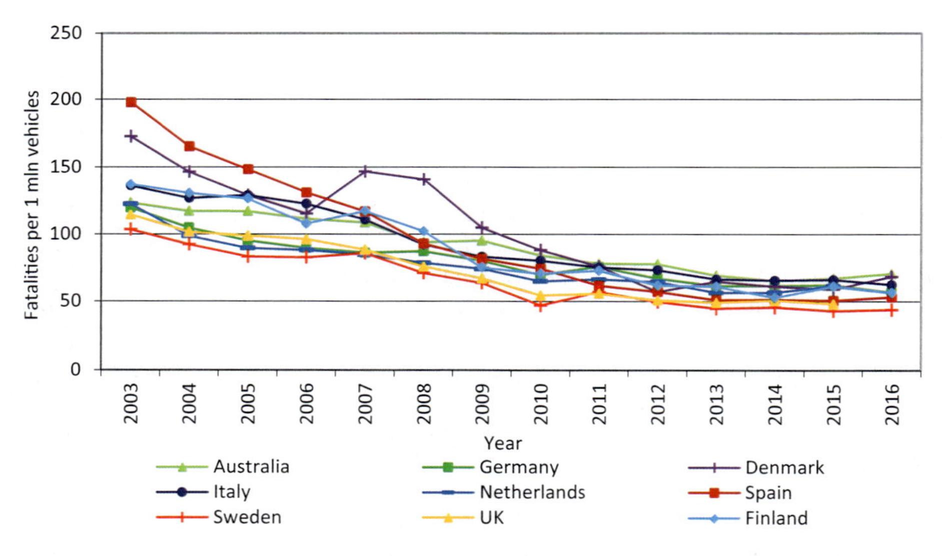

Fig. 3. Change in the number of fatalities in road accidents per 1 million vehicles in countries with high levels of traffic safety development (Data according to https://data.oecd.org/transport/road-accidents.htm#indicator-chart).

actual accident rates in developed countries lie below the Smeed's curve, while in Africa, Brazil, India, and China they are situated above the curve. However, the general pattern of the model is preserved in all countries [4]. Thus, with the growth of the country's motorization, a process of self-education of the nation takes place [6], as a result of which the accident rate may decrease even without taking special measures.

If we compare transport risks in the Republic of Belarus with the world trend according to Smeed, we get the following picture (Fig. 4).

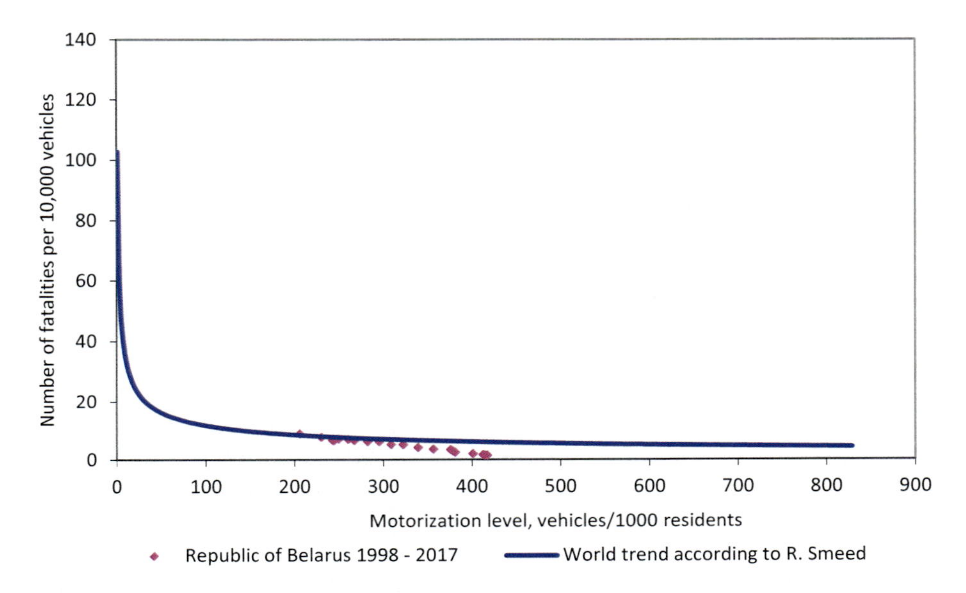

Fig. 4. Transport risks in the Republic of Belarus and the world trend (Data on the world trend according to [4]).

If we go back to the charts of Figs. 1 and 2, it can be seen that starting from 2011, the rate of reduction of accidents has been slowing down. Based on the situation up to the present time, transport risks and the number of fatalities can be approximated as follows (Fig. 5).

It can be assumed that the accident rate in the country asymptotically approaches the limit in accordance with Smeed's Law, and this limit is higher than in developed countries.

Remaining within the traditional for the Republic of Belarus engineering activity, which is related to road maintenance, it is unlikely that road traffic safety will be significantly improved. It is a well-known fact that after a certain level of motorization, the achievement of even lower values of transport and social risks will require a transition to qualitatively new practices and mechanisms of ensuring traffic safety.

Literature sources repeatedly emphasize that the limits for improving traffic safety are determined by the general possibilities of the road traffic safety system operating in the country. The system determines both the results and the steps to achieve them. Traffic safety limits for each country are limited by the institutional capacity. Therefore, many planned results may simply be technically unfeasible [7, 8]. As Bliss and Breen emphasize, without effective institutional governance, the country has little chance of implementing successful measures to achieve road traffic safety and the desired results [8].

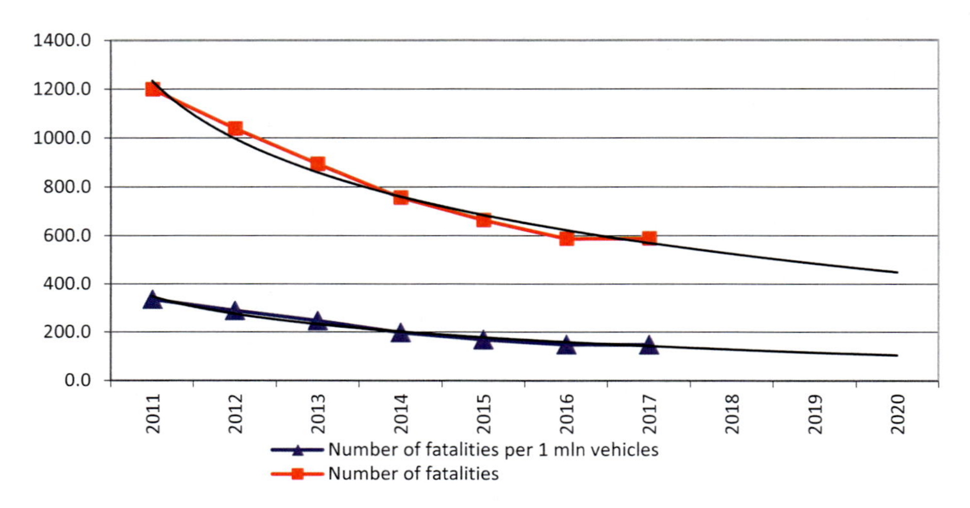

Fig. 5. Possible development of transport risks in the Republic of Belarus.

In such circumstances, it is logical to wish to use the experience of developed countries to achieve a lower accident rate.

4 Opportunities to Use the Best World Practices in Belarus

Developed countries have developed many systems and methods that can be used as an example. For instance, there are many countries that apply the Vision Zero system more or less successfully. We are carefully studying this experience and asking ourselves: will this experience help us?

Road safety concept, such as Vision Zero is extremely attractive for use because of its pronounced humanistic orientation. However, R. Elvik at the time stressed that even the application of ethical principles can be associated with moral dilemmas [9]. It is impossible to put the concept into operation by order. Despite the plain and simple principles of the concept, its application is associated with significant financial e expenses. Under the conditions of significant financial constraints, peculiar to developing countries, economic considerations prevail over ethical considerations underlying Road safety concept.

In our opinion, if we begin to apply this system today, we will not achieve many results. The main reason is that the people do not have the necessary understanding of the basic principles that underlie the system.

These are ethical principles, human peculiarities, responsibility, scientific data, as well as the interaction and interdependence of all components of the road and transport system. The Vision Zero system is based on the ethical principle of the unacceptability of deaths and serious injuries of people on the road [10]. The only acceptable number of fatalities and serious injuries is zero. The main idea of Vision Zero is based on understanding that the road and transport system must take into account the fact that humans are not perfect. Minor mistakes on the road too often lead to the death of people. Vision Zero explicitly states that the responsibility is shared by the system designers and the road user:

1. The designers of the system are always ultimately responsible for the design, operation and use of the road transport system and thereby responsible for the level of safety within the entire system.
2. Road users are responsible for following the rules for using the road transport system set by the system designers.
3. If road users fail to obey these rules due to lack of knowledge, acceptance or ability, or if injuries occur, the system designers are required to take necessary further steps to counteract people being killed or seriously injured [11].

It is these principles that must be widely promoted in our society today. For the time being, economic considerations prevail over ethical considerations when taking measures to improve traffic safety. For example, the value of life and injuries of people as a result of accidents are not defined in the Republic of Belarus for the purposes of assessing the effectiveness of traffic management measures and improving traffic safety.

An analysis of the situation in other countries that have been implementing road safety systems for a long time allows us to formulate some measures that will precede the introduction of Vision Zero for post-Soviet countries. This is a long-term program in mass media, aimed at changing the attitude of society to the problems of road accidents. Road designers and standards developers should introduce new design concepts, such as self-explaining and forgiving roads [12, 13].

There is another good experience in the European Union that we think deserves to be applied in Belarus. Namely, it is Directive 2008/96/EC "Road Infrastructure Safety Management."

The document introduces requirements and obliges the EU member states to introduce and implement the following activities on the roads of the Pan-European transport corridors [14]:

- traffic safety audit;
- certification of traffic safety auditors;
- assessment of the impact of road construction projects and road repairs on traffic safety;
- traffic safety inspection;
- road network safety rating (management of high accident concentration sections);
- informing the public about high accident concentration sections.

The Directive has been used effectively for almost 10 years to prevent road accidents in the EU and has the potential to be equally effective in other countries.

The introduction of a similar document has a great chance of success in the Republic of Belarus. This can be explained by the fact that the country has traditionally paid a lot of attention to the observance of technical standards and regulations. As of 2018, there are 49 technical standards in force in Belarus that are directly related to traffic safety. Requirements for traffic management facilities, their test methods and structure are contained in 41 documents. And if these standards are almost completely harmonized with similar European documents, the requirements for road infrastructure safety management processes are practically absent in Belarus. All technical standards are non-mandatory documents. Only technical regulations are mandatory in Belarus, which can be considered as an analogue of European Directives. Technical Regulation of the Customs Union TR/TS 014-2011 "Road Safety" is in force in relation to roads in Belarus. This document does not contain requirements for road infrastructure safety management processes as they are understood in European countries.

In such conditions, the transition to a qualitatively new level of work on improving road traffic safety requires the development of a national technical regulation on road infrastructure safety management.

5 Conclusion

Based on the foregoing, the following main conclusions can be made.

The implementation of the Road Traffic Safety Concept in the Republic of Belarus has yielded positive results in the form of reduction of the number of road accidents and fatalities and injuries in them. Further reduction of accidents requires the introduction of new practices and mechanisms.

The experience of developed countries where the accident rate is low shows that without effective institutional management, the country has little chance of implementing successful measures to improve road safety and achieve the desired results in this area.

Efficient road infrastructure safety management measures, such as traffic safety audits, traffic safety inspections, and management of high accident concentration sections, have been effective in preventing traffic accidents in the EU for almost 10 years and have the potential to be equally effective in other countries.

The implementation of road infrastructure safety management processes in the Republic of Belarus is currently hampered by the lack of the necessary regulatory framework. To remedy the situation, it is required to develop the national technical regulation on road infrastructure safety management and a set of related standards of the Republic of Belarus.

Ignoring these facts will lead to the Republic of Belarus progressively falling behind developed countries, first of all the EU countries, in the issues of road traffic safety improvement.

References

1. Global Status Report on Road Safety 2013: Supporting a decade of action. apps.who.int/iris/bitstream/10665/78256/1/9789241564564_eng.pdf. Accessed 30 Nov 2018
2. Road Safety – Considerations in Support of the 2030: Agenda for sustainable development. https://unctad.org/en/PublicationsLibrary/dtltlb2017d4_en.pdf. Accessed 30 Nov 2018
3. Road Safety Annual Report 2018. http://www.unece.org/fileadmin/DAM/trans/doc//wp1/Improving_Global_Roady_Safety_2011.pdf. Accessed 10 Nov 2018
4. Koren, C., Borsos, A.: Is Smeed's Law still valid? A world-wide analysis of the trends in fatality rates. J. Soc. Transp. Traffic Stud. **1**, 64–76 (2010)

5. Smeed, R.J.: Some statistical aspects of road safety research. J. R. Stat. **1**(112), 1–34 (1949)
6. Adams, J.G.U.: Smeed's Law: some further thoughts. Traffic Eng. Control **2**(28), 70–73 (1987)
7. Towards Zero: Ambitious road safety targets and the safe system approach. https://fevr.org/wp-content/uploads/2017/12/Towards-Zero-OECD-PDF-57-MB.pdf. Accessed 10 Nov 2018
8. Bliss, T., Breen, J.: Implementing the recommendations of the world report on road traffic injury prevention. In: Country Guidelines for the Conduct of Road Safety Capacity Reviews and the Related Specification of Lead Agency Reforms, Investment Strategies and Safety Projects. http://documents.worldbank.org/curated/en/712181469672173381/pdf/815980WP0Traff00Box379836B00PUBLIC0.pdf. Accessed 10 Nov 2018
9. Elvik, R.: Can injury prevention efforts go too far? Reflections on some possible implications of Vision Zero for road accident fatalities. Accid. Anal. Prev. **31**, 265–286 (1999)
10. Johansson, R.: Vision Zero – implementing a policy for traffic safety. Saf. Sci. **47**, 826–831 (2009)
11. Larsson, P., Dekker, S.W.A., Tingvall, C.: The need for a systems theory approach to road safety. Saf. Sci. **48**, 1167–1174 (2010)
12. La Torre, F., Saleh, P., Cesolini, E., Goyat, Y.: Improving roadside design to forgive human errors. Procedia – Soc. Behav. Sci. **53**, 235–244 (2012)
13. Theeuwes, J., Godthelp, H.: Self-explaining roads. Saf. Sci. **19**(2–3), 217–225 (1995)
14. European Directive on Road Safety Management [2008/96/EC]: Guidelines for competent authorities on the application of the directive. https://www.gov.uk/government/uploads/system/uploads/attachment_data/file/3565/guidelines.pdf. Accessed 6 Aug 2018

Experimental Research on Shock Absorbers of Light Vehicles

Saulius Nagurnas(✉) and Paulius Skačkauskas

Vilnius Gediminas Technical University, J. Basanavičiaus G. 28, 03224 Vilnius, Lithuania
{saulius.nagurnas,paulius.skackauskas}@vgtu.lt

Abstract. Zero fatalities in road transport and overall traffic safety is one of the most important common goals of the society, industry and academia. To achieve this goal, various researches in different fields, from the analysis of the materials and designs to the development of autonomous ground vehicles, are being done. Respectively, vehicle suspension and shock absorbers are essential elements, which significantly contribute to the oscillations of sprung and unsprung masses, also to the handling and braking characteristics of a vehicle, i.e. they affect the safety of the vehicle. Due to this reason, in this work, an experimental research on the condition of the shock absorbers of light vehicles, when applying the original research methodology, is presented. The performed experimental research is based on two stages. Firstly, the condition of the shock absorbers of the researched vehicles was determined by using a shock absorber test stand. Secondly, equipment was installed in the test vehicles, which measures the vertical acceleration values of the vehicle body, when it drives over an obstacle at low velocities. Based on the performed analysis of the experimental results, vertical acceleration values of the sprung masses of vehicles were indicated, which allows to simply and effectively evaluate the condition of the vehicle shock absorber, without a complex investigation of the shock absorber.

Keywords: Traffic safety · Light vehicle · Shock absorber · Vertical acceleration

1 Introduction and Literature Review

Currently many vehicle manufacturers focus on increasing the safety, comfort and control of the vehicles. These properties are majorly influenced by the suspension of the vehicle and one of its main components – shock absorbers. The suspension connects the frame or the body of the vehicle with the wheels, reduces the impact from road irregularities, and dampens the vibrations. This way the traffic safety and the comfort of driving are increased, the passengers and the vehicle are protected from unnecessary overload that can appear during oscillations. That is why, it is essential to continually improve the optimal performance parameters of the vehicle suspensions, considering the condition of one of the most important elements – shock absorbers.

Shock absorbers that are well designed and have proper damping efficiency should minimize the vibrations of the vehicle bodywork and ensure the appropriate road

A. Varhelyi et al. (Eds.): VISZERO 2018, LNITI, pp. 120–129, 2020.
https://doi.org/10.1007/978-3-030-22375-5_14

surface grip of the wheels [1]. It is clear that shock absorbers, which do not meet these essential requirements, directly endanger traffic safety by negatively affecting the driver of the vehicle and the performance of different vehicle components. A number of research works can be found, which analyse the relation and interaction between the shock absorbers, other vehicle components, the driver and traffic safety. For example, the authors in [2] investigated the rattling noise of shock absorbers, which substantially affects the psychological and physiological perceptions of the driver. The influence of a shock absorber wearing on the vehicle braking performance was analysed in [3]. The authors noted that the shock absorber wearing status has no influence on the brake performance of the vehicle when the car is on a smooth road profile, however, in the case of a rough road profile, the shock absorber status has a significant negative influence on the stopping distance. The influence of worn shock absorbers on anti-lock brake systems was also investigated in [4]. The authors in [5] considered the influence of the shock absorber on the steerability of the vehicle motion. The results of their investigation showed that the damaged shock absorber actually causes the change of the profile steerability of the vehicle and increases vehicle understeering. The change of the vehicle ride comfort and road holding characteristics due to a change of shock absorber temperature was examined in [6]. In their research, the authors remarked that the decrease in temperature does influence the ride comfort of vehicles, expressed in growth in spectral density of vertical acceleration. Other researches, which similarly to [6], are based on the influence analysis of the specific parameters of the shock absorbers on ride comfort /safety, like [7, 8], also can be found. In [7] the authors considered the influence of shock absorber friction on vehicle ride comfort and, in [8], the influence of shock absorber installation angle on vehicle handling and ride quality was studied. Based on the above mentioned research works it can be stated that the influence of shock absorbers on traffic safety and proper evaluation of the shock absorber condition is a relevant problem. Respectively, different propositions for evaluating the technical condition of shock absorbers can be found in literature. A review work on most common experimental diagnostic assessments of the technical condition of the shock absorbers is given in [9]. The authors pointed out three basic experimental diagnostic approaches: testing the shock absorbers using the free vibration method, testing the shock absorbers using the forced vibration method and the damping coefficient measurement. However, in their work, the authors do not discuss in detail the advantages and disadvantages of the mentioned approaches. The authors in [10] used an assessment of the technical condition of the vehicle shock absorbers built into the vehicle on the basis of the signals recorded during the vibration test. Based on their research, the authors also proved that a change of oil leak in the shock absorber causes a negative effect and increases the vertical accelerations amplitude of the wheel. An attempt to diagnose the damage of the shock absorbers of a light vehicle during road operation with the use of a vibration response measurement has been described in [11]. In the work it is indicated that the described method allows to find not only the fact that there is a problem with a shock absorber, but even which particular shock absorber is faulty. A novel approach for monitoring the condition of shock absorbers was proposed in [12]. In the novel approach the authors suggested that instead of monitoring the change in dynamic performance of a suspension system, the age and condition of a shock absorber can be estimated by measuring the cumulative work done using a calorimetry method involving temperature sensors. In [13] it is stated that due to the non-linear dynamics of shock absorbers and the strong influence of the disturbances such as the road profile, fault detection of a semi-active shock absorber is a challenge. Thus, in [13], two model-based fault algorithms for a semi-active shock absorber were proposed and compared: an observer-based approach and a parameter identification approach. While taking into consideration the researches described in different sources, it can be seen that shock absorbers are one of the most important vehicle subassemblies that significantly affect such parameters as braking effectiveness, stability, road holding and ride comfort characteristics, etc. Thus, diagnostics of such elements is a major factor which can ensure better traffic safety. However, as can be noted from the literature review, the diagnostic process of the shock absorbers itself can be very complex and requires specific knowledge. Due to these reasons, the aim of this work is to propose a simpler and effective methodology based on the evaluation of average values of the vertical acceleration a_y of the sprung masses, which would allow to indicate the performance and condition of the shock absorber without a complex investigation.

The remainder of this work is organised as follows. In Sect. 2, the experimental procedure is described. Section 3 provides the results and discussion of the experimental procedure. The final section presents the conclusions of the work.

2 Experimental Research

To achieve the aim of the work, an experimental research, during which the technical condition of the front and rear axle shock absorbers of the test vehicles was determined, was performed. The experimental research was performed in two stages. Firstly, the technical condition of the shock absorbers of the test vehicles was determined under laboratory conditions using a shock absorbers test stand. Secondly, to imitate real life vehicle exploitation conditions, the experimental research was performed in an open public road, while the test vehicles were moving over a speed bump. In total, the experimental researches were performed with 10 test vehicles.

2.1 Research Under Laboratory Conditions

During the first stage, the experimental research was performed in order to estimate which test vehicles will be used in further experimental tests. This stage of the experimental research is important in order to precisely indicate vertical acceleration values which will allow to evaluate the condition of the vehicle shock absorbers, i.e., to eliminate the vehicles that have shock absorbers with poor damping efficiency, which would have an impact on the formation of errors. In this case, testing of the shock absorbers, while applying the forced vibration method, was performed.

As already noted, the experimental research was performed using a *SAFELANE 400/800* shock absorbers diagnostics stand, designed to diagnose the shock absorbers damping efficiency indicators. In general, the used diagnostics stand consists of two measuring plates which consistently transfer to each other oscillations of variable frequency and constant amplitude. The stand measures the vertical force that is transferred from the wheel to the measuring plate. During the measuring, it is sought to determine the *Eusama* value, which shows the ability of the suspension elements to ensure the road surface grip of the wheels under unfavourable traffic conditions, and in this way to check the shock absorbers damping efficiency. For proper understanding of the shock absorbers damping efficiency evaluation process, it must be pointed out that the *Eusama* value is the minimal vertical resonant force to the wheels, which is expressed as a percentage of the wheel static mass. The evaluated *Eusama* value can change from 0 to 100% and respectively, the determined values can be divided into: from 0 to 20% – poor damping efficiency, 21 to 40% – sufficient damping efficiency, 41 to 60% – good damping efficiency, above 60% – excellent damping efficiency. An example of the experimental research results is given in Fig. 1.

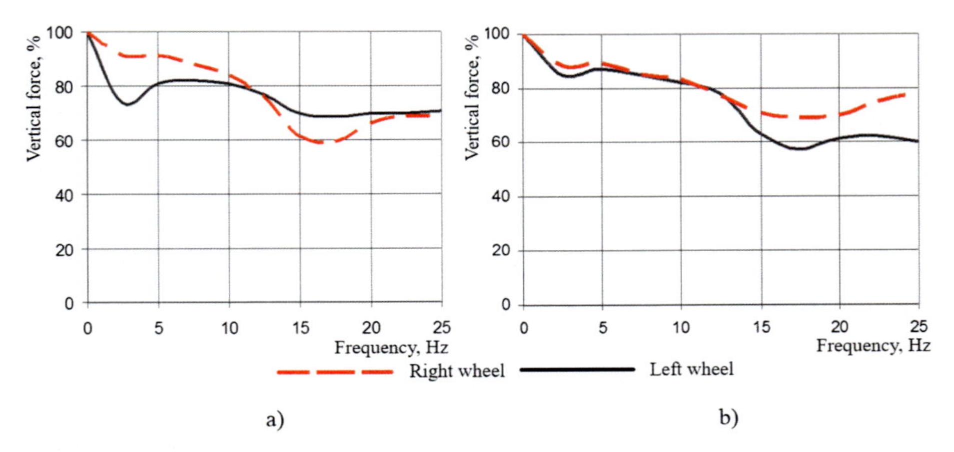

Fig. 1. The influence of the oscillations frequency of the front and rear axles on the value of the vertical forces affecting the wheels: (a) – front axle of test vehicle No. 1; (b) – rear axle of test vehicle No. 1 (Source: by the authors).

In the presented results in Fig. 1, the horizontal coordinate shows the vibration frequency of the measuring plates of the used stand, the range of which is from 0 to 25 Hz. The vertical coordinate, respectively, shows the relative value of the vertical force, which is affecting the wheel and is transferred to the measuring plates. It also must be remarked that, in Fig. 1, the curves show the minimal vertical force relative value of the vibrating wheel that is transferred to the measuring plates at different vibration frequency. Thus, the higher the curve, the smaller the change in the vertical force, the better the shock absorbers damping efficiency. As a result of this stage of the experimental research, it is important to note that the *Eusama* values of all the tested vehicles were within the value ranges corresponding to good and excellent damping of the shock absorber, i.e., these values of all the vehicles were within 41–60% and above 61%.

While taking into consideration the goal of the first stage of the experimental research, based on the results, after testing the shock absorbers damping efficiency of all the test vehicles using the diagnostics stand it can be stated that all of these vehicles can be used in further research, while moving in an open public road and imitating real life vehicle exploitation conditions.

2.2 Experimental Research While Moving in an Open Public Road

During the second stage, the experimental research was performed in order to estimate the value of the test vehicles sprung masses oscillations in the vertical direction, when the vehicle is moving over a speed bump and, based on the estimated values, to find out at which values of the vertical acceleration it can be stated that the shock absorbers in vehicles have proper damping efficiency (ensuring safe control of the vehicles). To achieve this goal, measuring equipment was installed in all of the test vehicles which is designed to evaluate the vertical acceleration a_y of the oscillations of the vehicle body as accurately as possible:

- data collection /processing device *KISTLER DAS-3* (was installed in the interior of the test vehicles (Fig. 2, part a)). The device was used to systemise and process the data measured during the experimental tests;
- dynamic angular deviation and acceleration sensor *KISTLER TANS-3215003M5* (was installed in the interior of the test vehicles (Fig. 2, part a)). The device was used to measure the oscillations and acceleration of the vehicle body;
- diagnostic equipment *BOSCH KTS570* (was installed in the interior of the test vehicles). The device was used for accurate measuring of the test vehicle velocities during experimental drives.

a) b)

Fig. 2. Second stage of the experimental research: (a) – vertical acceleration sensors with the data collection /processing system; (b) – the open public road with the speed bump where the experimental research was performed (Source: by the authors).

The tests were done on an asphalt-concrete coat in an open public road driving over a speed bump (Fig. 2, part b). The speed bump in the shape of a part of a circle in the street was set up over the entire width of the roadway. The width of the bump was 0.9 m, and the height – 0.05 m. What is more, during the second stage of the experimental research, different experimental drives were performed, i.e., the tests were done at the velocities of 10 km/h, 20 km/h and 30 km/h. During all the experimental drives, in all of the tested vehicles there were two human supervisors: the driver and the passenger.

3 Results and Discussion

In general case, the change patterns of the vertical acceleration values a_y, depending on time and the position of the test vehicle in relation to the speed bump, can be explained by the dependency presented in Fig. 3.

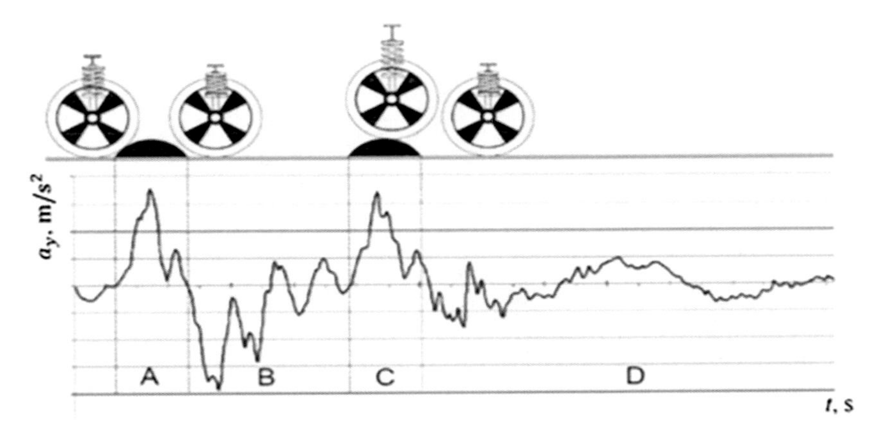

Fig. 3. The change patterns of the vertical acceleration values a_y, when the vehicle drives over a speed bump (Source: by the authors).

Figure 3 can be explained as follows: at the top of the graph portrayed is the wheel of the vehicle with a shock absorber and a spring at the moment of driving when the vehicle drives over the speed bump with the front and rear wheels. The condition of the spring (compression, rebound) and respectively, the compression and rebound of the shock absorber are portrayed, when the wheels drive on and off the speed bump. In Fig. 3, part A reflects the moment and the values of the vertical acceleration a_y, when the front wheels of the vehicle drive over the speed bump. The highest values of the appearing acceleration a_y are noticed at this moment. Part B indicates the acceleration values of the sprung masses of the vehicle, when the front wheels of the vehicle make contact with the road surface after driving off the bump. At this moment, the damping of the oscillation of the sprung masses and the high growth of the vertical acceleration values a_y can be noticed. In part C, portrayed is the moment when the vehicle drives on and off the speed bump with the rear wheels and the highest acceleration a_y values are seen. Part D indicates the vertical acceleration values of the sprung masses of the vehicle, when the rear wheels of the vehicle make contact with the road surface after driving off the bump.

Actual examples of the results of the second stage of the experimental research are respectively given in Figs. 4 and 5. It must be noted that the results of the experimental drives of all the test vehicles were similar to each other. Thus, only a few specific examples of the experimental data recorded during the experimental drives are described further.

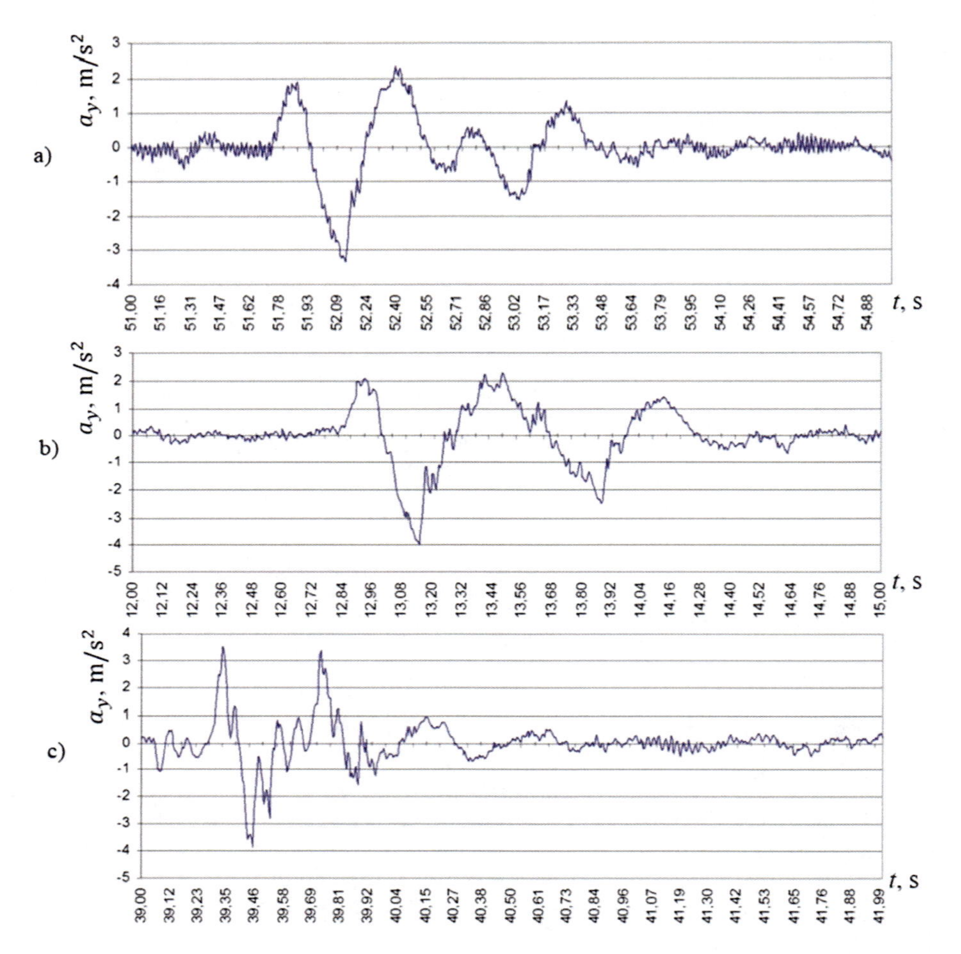

Fig. 4. The change patterns of the vertical acceleration a_y, when the test vehicle No. 1 drives over the speed bump: (a) – at the velocity of 10 km/h; (b) – at the velocity of 20 km/h; (c) – at the velocity of 30 km/h (Source: by the authors).

In Fig. 4 an example of the change pattern of the vertical acceleration a_y of test vehicle No. 1, while driving over the speed bump at different, afore mentioned, velocities, is presented. From all of the cases of Fig. 4, the damping of the vertical oscillations of the vehicle sprung masses, when the wheels of the vehicle drive off the bump, can be clearly seen. When the vehicle drives over the bump with the front wheels, the value of the acceleration in the vertical direction a_y grows exponentially and then starts to gradually decrease. When the vehicle drives over the bump with the rear wheels, the value of the acceleration a_y increases and starts to decrease equally quickly, and then settles down during smooth driving. Respectively, based on the data of all the experimental drives, it can be indicated that the highest values of the vertical acceleration a_y, when the acceleration measuring sensors are installed on the front part of the vehicle (on the front glass of the vehicle), depending on the vehicle velocity, when driving on the speed bump, varies from 0.6 m/s^2 to 5.25 m/s^2. This observation can be explained by the above given figure (Fig. 3). In Fig. 5, the test results of three test vehicles, in the figure respectively marked as 1, 2 and 3, are also presented.

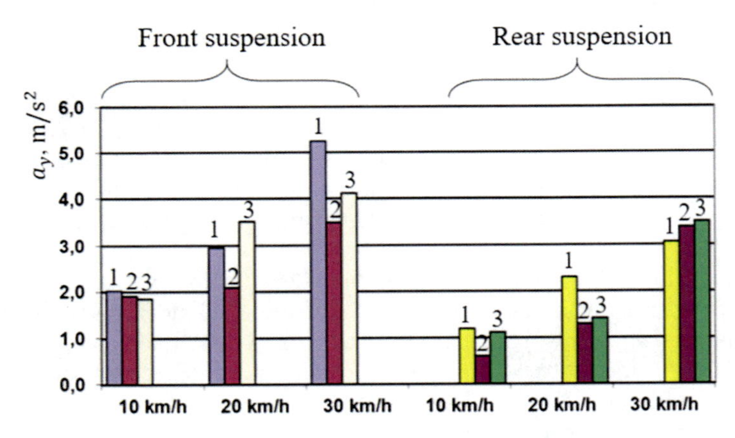

Fig. 5. Example of different test vehicles front and rear axle vertical acceleration a_y values, depending on the driving velocity (Source: by the authors).

Based on the results given in Fig. 5 it can be pointed out that, when driving over an obstacle, the average values of the vertical acceleration a_y of the sprung masses of the vehicle, when the vehicle has proper and efficient shock absorbers, are around 1.93 m/s^2 when driving with the front wheels and 0.97 m/s^2 when driving with the rear wheels. Respectively, at the velocity of 20 km/h – around 2.85 m/s^2 and 1.67 m/s^2, at the velocity of 30 km/h – around 4.30 m/s^2 and 3.32 m/s^2. It can be seen that, in all the cases of the experimental drives, the highest acceleration values were reached when driving over the speed bump with the front wheels of the vehicle. Respectively, during all the cases of the experimental drives, the highest acceleration values are reached at the highest, i.e., 30 km/h, velocity. Thus, these average values of the vertical acceleration a_y can be used as reference values seeking to evaluate whether the shock absorbers can ensure safe exploitation of the vehicle, without a detailed inspection of the shock absorbers rebound and compression performance, which requires explicit knowledge about their dynamic behaviour under various operational modes.

4 Conclusions

In this work an experimental procedure of shock absorbers condition evaluation was presented and described. The described experimental procedure is based on an approach combined of two stages: firstly, testing of the shock absorbers, while applying the forced vibration method, and secondly, an experimental research while moving in an open public road, were performed. Main contribution of this work – indicated vertical acceleration values of the vehicle sprung masses, which allows to simply and effectively evaluate the condition /fault of the vehicle shock absorbers, without a complex investigation of the shock absorber or specific explicit knowledge. The analysis of the experimental research showed that the approximate average values of the vertical acceleration of the sprung masses, at which an assumption can be made that the shock absorbers of the vehicle function properly, are: at the velocity of 10 km/h – 1.93 m/s^2 when driving with the front wheels and 0.97 m/s^2 when driving with the rear wheels, at the velocity of 20 km/h, respectively – around 2.85 m/s^2 and 1.67 m/s^2, at the velocity of 30 km/h – around 4.30 m/s^2 and 3.32 m/s^2. Respectively, practical application possibilities of the described results can be pointed out – the determined values of the vertical acceleration of the sprung masses can be used as guidelines of shock absorber and suspension condition, while performing the periodic motor vehicle inspection.

However, although the combination of the two different experimental procedure approaches increased the accuracy of the performed research, to improve the proposed assumption and values, it would be appropriate to carry out the research with a greater number of light vehicles. Due to this reason, future work by the authors is to further perform experimental research, while considering not only a greater number of test vehicles, but also a different size of obstacles and respectively, different velocities.

References

1. Luczko, J., Ferdek, U.: Non-linear analysis of a quarter-car model with stroke-dependent twin-tube shock absorber. Mech. Syst. Signal Process. **115**, 450–468 (2019)
2. Huang, H.B., Li, R.X., Huang, X.R., Yang, M.L., Ding, W.P.: Sound quality evaluation of vehicle suspension shock absorber rattling noise based on the Wigner-Ville distribution. Appl. Acoust. **100**, 18–25 (2015)
3. Calvo, J.A., Diaz, V., San Roman, J.L., Garcia-Pozuelo, D.: Influence of shock absorber wearing on vehicle brake performance. Int. J. Automot. Technol. **9**(4), 467–472 (2008)
4. Koylu, H., Cinar, A.: The influences of worn shock absorber on ABS braking performance on rough road. Int. J. Veh. Des. **57**(1), 84–101 (2011)
5. Parczewski, K.: Exploration of the shock-absorber damage influence on the steerability and stability of the car motion. J. KONES Powertrain and Transp. **18**(3), 1–8 (2011)
6. Pavlov, N.: Influence of shock absorber temperature on vehicle ride comfort and road holding. In: 9th International Scientific Conference on Aeronautics, Automotive and Railway Engineering and Technologies, pp. 1–6. MATEC Web Conference, Bulgaria (2017)
7. Fujimoto, G., Komori, K., Tsukamoto, T., Sogawa, N., Nishimura, T.: Influence of shock absorber friction on vehicle ride-comfort studied by numerical simulation using classical single wheel model. SAE Technical Paper 2018-01-0692, pp. 1–5 (2018)

8. Pananda, N., Daowiangkan, S., Intaphrom, N., Kantapam, W.: Influence of shock absorber installation angle to automotive vehicle response. J. Res. Appl. Mech. Eng. **5**(2), 94–105 (2017)
9. Buczaj, M., Walusiak, S., Pietrzyk, W.: Diagnostic assessment of technical condition of the shock absorbers in automotive vehicles in a selected diagnostic station. In: Commission of Motorization and Power Industry in Agriculture, pp. 59–66. Commission of Motorization and Power Industry in Agriculture, Poland (2007)
10. Konieczny, L., Burzdzik, R., Lazarz, B.: Application of the vibration test in the evaluation of the technical condition of shock absorbers built into the vehicle. J. VibroEng. **15**(4), 2042–2048 (2013)
11. Bialkowski, P., Krezel, B.: Diagnostic of shock absorbers during road test with the use of vibration FFT and cross-spectrum analysis. Diagnostyka **18**(1), 79–86 (2017)
12. Howard, C.Q., Sergiienko, N., Gallasch, G.: Monitoring the age of vehicle shock absorbers. In: International Conference on Science and Innovation for Land Power, pp. 1–5. Industry Defence and Security, Australia (2018)
13. Hernandez-Alcantra, D., Morales-Menendez, R., Amezquita-Brooks, L.: Fault detection for automotive shock absorber. J. Phys: Conf. Ser. **659**, 1–12 (2015)

How Congruent Can Human Attitudes, Intentions and Behaviour Be: The Case of Risky Driving Behaviour Among Lithuanian Novice Drivers

Laura Šeibokaitė[1(✉)], Justina Slavinskienė[1], Renata Arlauskienė[2,3], Auksė Endriulaitienė[1], Rasa Markšaitytė[1], and Kristina Žardeckaitė-Matulaitienė[1]

[1] Vytautas Magnus University, Kaunas, Lithuania
laura.seibokaite@vdu.lt
[2] Klaipeda University, Klaipeda, Lithuania
[3] Klaipeda State University of Applied Sciences, Klaipeda, Lithuania

Abstract. The aim of this paper is to evaluate the possibilities to predict risky driving behaviour of novice Lithuanian drivers during the first nine months of their independent driving by risky attitudes and intentions to risk assessed before they were licenced to drive.

188 novice drivers participated in a three-wave longitudinal study. They were approached at driving schools in Lithuania in the beginning of driving training and after it and asked to fill in the questionnaire. To assess the risky driving participants were interviewed one year after the end of driving training.

Driving errors correlated with attitudes and intentions only for female drivers. Violations while driving were related to attitudinal variables for both males and females. Violations as well as driving errors could be predicted by attitudes and intentions measured at the same time as behaviour, when variables of other measurements were controlled.

Risky driving of drivers could be better explained by their attitudes and intentions, when they have already gained some driving experience. It might be presumed that driving experience shape risky attitudes and intentions, not vice versa as it was expected by theory.

Keywords: Risky driving · Risky attitudes · Intentions · Novice drivers

1 Introduction

Risky driving and the prevalence of its negative consequences remain as one of the major problems all over the world [1, 2]. Based on statistics of European Commission CARE database, Lithuania exceeded European average number of deaths during road accidents (66 people for 1 million population in Lithuania while in ES – 52 people) [3]. According to data of the Lithuanian Road Police Office (2017) almost a half of all recorded traffic accidents within all groups of drivers was caused by inexperienced drivers (those who have less than 2 years driving experience). Comparison of

A. Varhelyi et al. (Eds.): VISZERO 2018, LNITI, pp. 130–139, 2020.
https://doi.org/10.1007/978-3-030-22375-5_15

statistics in 2017 and 2018 revealed that each year more novice drivers make serious traffic offences in the first two years of independent driving. For example, in 2018 287 novice drivers exceeded speed limit more than 30 km/h while in 2017 the number was 250. In 2018, 134 novice drivers drove under the influence of alcohol (0.4–1.5 BAC) while in 2017 it was 123 novice drivers [4]. Thus, risky driving among novice drivers still is one of serious issue in traffic safety.

Previous research confirmed that the first year of independent driving is a crucial period for road traffic rules violations and accidents [5]. Young novice drivers are at higher risk while driving because of insufficient driving skills and willingness to take a risk while driving, make impression on peers etc. [6–9]. It was found that maturation contribute significantly to the decrease of risky driving [10]. Thus, more extensive focus on psychological factors related to novice drivers' risky driving is crucial to understand the development of risky driving patterns at early stages [11] as well as to make significant improvements in driving training programs [12].

Theory of Planned Behaviour [13] was created to explain the relationships among attitudes, intentions, and reasoned human behaviour in various areas of functioning [13]. Ajzen described *attitudes* as positive or negative evaluation of certain behaviour. *Intentions* could be identified as the willingness and readiness to put the efforts to perform certain action. Positive attitudes form the intentions to act in certain way, therefore intentions lead to that specific behaviour [13]. Ajzen introduced few other constructs as perceived behavioural control and perceived social norms as well [13], but they are out of the scope for this paper, they are not introduced further. It should be noted that 'a behavioral intention can find expression in behavior only if the behavior in question is under volitional control, i.e., if the person can decide at will to perform or not perform the behavior' [14, pp. 181–182]. There might be arguing if driving is the type of behaviour that person rationally plan each time when performing [14]. Indeed, it might be treated as automatized behaviour or habit [15].

There were numerous attempts to explain risky driving behaviour by TPB [16–18]. Previous research found that more favourable attitudes towards risky driving (over the speeding, drink driving, phone use while driving, fatigued driving and riding with intoxicated driver) were positively related to intentions to behave in particular way [18–20]. Full model of TPB was tested and the results revealed that attitudes towards risky driving explained intentions to drive in a risky manner which explained self-reported risky driving in motorcyclists and truck drivers [17, 21].

The results allowed presuming that attitudes, intentions and driving behaviour could be congruent [22]. In other words, usually drivers drive in the way they initially intended and believed it is appropriate. For this reason, sometimes, researchers assess attitudes and intentions instead of behaviour when it cannot be observed directly [19]. In driving research context, the attitudes as behavioural indicators were assessed in pre-drivers when young people still did not have driving experience and were unable to report their driving peculiarities [23]. However, some studies failed to find significant relationships among TPB constructs. For example, Paris and Van den Broucke [24] reported that intentions to obey speed limits were not related to attitudes towards speeding behaviour; objective speeding behaviour and number of self-reported violations were not explained by intentions related to certain risky behaviour [24]. Therefore, behavioural variables should be substituted by attitudinal constructs cautiously.

The prediction of future driving behaviour based on attitudes and intentions to drive in a risky manner is even more problematic. Previous longitudinal studies revealed that intentional violations of novice drivers were predicted by more favourable attitudes towards speeding, measured before gaining the driving license. However, attitudes failed to predict later driving errors in the same sample [25]. Similar results were found by Rowe et al. [12]. Attitudes towards to speeding of non-drivers were not significant in predicting future driving behaviour as they became fully-qualified drivers [12].

Still, it is questionable, if pre-driving attitudes and intentions are valid indicators of later risky driving. While pre-drivers have no experience of independent driving, they cannot develop the realistic representation of own driving style, despite the initial attitudes and intentions related to driving [26]. Even though TPB proposed and empirical studies supported the idea that attitudes form intentions and later behaviour [13, 22], reciprocal relationship between attitudinal and behaviour characteristics might be expected as well [27]. Results of some studies might serve as evidence of the assumption. Forward [16] found that self-reported intention to overtake dangerously was related to the frequency of this behaviour in the past [16]. Attitudes towards speeding became riskier at the beginning of their solo driving in comparison to driving training. However, other attitudes towards risky driving (close following and overtaking) became safer over this period [28]. It could be presumed that real driving experience changes the way how novice drivers perceive appropriate behaviour on the road.

Despite the results of previous studies, there is a lack of evidence how pre-driving attitudes and intentions can predict behaviour on the road while gaining some driving experience. Therefore, the aim of this paper is to evaluate the possibilities to predict risky driving behaviour of novice Lithuanian drivers during the first nine months of their independent driving by safety attitudes and intentions to risk assessed before they were licenced to drive.

2 Method

2.1 Subjects

In total, 188 novice drivers (71 males and 117 females) participated in three study stages (see Sect. 2.2). The average age among males were 19.52 (±3.64) years (ranged from 16 to 31 years), while in females' group – 21.58 (±6.65) years (ranged from 17 to 53 years). At the first and the third stages of study, participants were asked about their driving experience. Analysis showed that one third (n = 63, 33.5%) of all driving license candidates had no driving experience when they started driving training course. However, 32.4% of them (n = 61), reported that they had some driving experience (driving with other experienced driver). At the third stage, a half of all participants (n = 103, 54.8%) drove on daily basis or 4–6 times per week. Comparison analysis of driving experience showed that novice driver males had similar driving experience as novice driver females (χ^2 (5) = 8.65, p = .12).

2.2 Procedure

The longitudinal design was applied and consisted of three stages. Firstly, driving license candidates who signed for driving training courses were invited to participate and fill out the self-reported questionnaire during the theory classes. The same driving license candidates were surveyed at the end of driving training courses (after 2 months), when they passed theoretical exams at driving school. The third stage was carried out after nine months from the second stage of the study. Those novice drivers who participated in the earlier stages and who drive independently more than a month were asked to fill out the same self-reported questionnaire online or via personal email. Participants were asked to participate on voluntary basis. They were informed about the three-stage study and other ethical issues in written form.

2.3 Instruments

The Driver Behavior Questionnaire (DBQ) [29] was used to examine self-reported risky driving. 24-item questionnaire scored on a 5-point Likert scale ranging from 1 – "strongly disagree" to 5 – "strongly agree". Two aspects of risky driving: the driving errors (16 items) and intentional traffic rules violations (8 items) were evaluated. Originally this questionnaire measures three types of risky driving behaviour: violations, errors, and lapses, but a two-factor solution fit the data better in Lithuanian drivers' sample [30]. Driving errors consisted of actions that are not planned and appeared mostly because of some mistakes (e.g. "Misjudge your crossing interval when turning right and narrowly miss collision") while intentional traffic rules violations (e.g. "Get involved in unofficial "races" with other car drivers") were considered to be deliberate deviations from safe driving practices. The higher scores in each scale indicated riskier driving – more driving errors and more intentional violations. The internal validity of each scale was sufficient: Cronbach α ranged from .75 (for intentional violations) to .85 (for driving errors). Driving behaviour was measured in the third (T3) stage of the study.

In order to determine the attitudes of novice drivers towards risky driving, the questionnaire *Attitudes towards Risky Driving* was used [31]. The questionnaire (16 statements) measured several aspects of the attitude towards traffic safety: attitude towards non-compliance with the rules and over speeding, attitude towards risky driving of other persons and attitude towards drink-driving (e.g. "If you are a good driver, it is acceptable to drive a little faster"; "I would never drive after drinking alcohol"). Attitudes towards risky driving were measured by Likert scale from 1 – "strongly disagree" to 5 – "strongly agree". Higher score of attitudes towards risky driving indicated the risk-favourable attitude towards driving. The internal validity of the scale in all three measurements was sufficient – Cronbach α ranged from .76 to .82. The answers of the first (T1), second (T2) and third (T3) stages of the study were used in further analysis.

Intentions to drive in risky manner were measured by four scenarios related to safe distance keeping, drunk driving, speeding, dangerous overtaking [32]. Each outline included a line drawing together with a short description (see [32], p. 95). Respondents were instructed to imagine themselves in the depicted set of circumstances and make judgments on each standardized and hypothetical scenarios by Likert scale from 1 "Never" to 7 "I always would do this". Higher scores indicated stronger intentions to drive in a risky manner and stronger intentions to violate road traffic rules. The internal validity of the scale was sufficient – Cronbach α = .64. Data about intentions to drive in risky manner was collected in the second (T2) and the third stages (T3) of the study.

Demographic data was obtained, and it included gender, age and driving experience.

3 Results

3.1 Relationship Among Safety Attitudes, Intentions, and Driving Behaviour

To assess relationships among attitudes towards road safety, intentions to drive in a risky manner, and driving behaviour (errors and violations) the correlational analysis using Spearman's coefficient was performed. Results were presented for males and females separately as scholars revealed gender differences in risky driving [33].

Data showed that driving errors in male drivers were significantly correlated only with intentions to drive risky measured at the same time as behaviour (Table 1). Violations while driving were positively related to risk favourable attitudes before training and after some driving experience, intentions to drive risky at the latest measurement as well. Safety attitudes and intentions measured right after training were not related to later behaviour on the road for male novice drivers. It should be noted that attitudes and intentions measured right after training did not correlated with attitudes and intentions nine months later. Attitudes towards risky driving correlated positively with risk-taking intentions at the same measurement time, but coefficients were rather moderate.

Table 1. Correlations among driving behaviour, attitudes, and intentions in male drivers.

	Violations	Attit.1	Attit.2	Inten.2	Attit.3	Inten.3
Errors	.208	−.067	−.148	−.163	.119	.316**
Violations		.336***	.096	.128	.489***	.432***
Attit.1			.311***	.212**	.260*	.387***
Attit.2				.473***	−.030	−.082
Inten.2					−.044	.074
Attit.3						.360***

Note: Attit.1 = Attitudes at T1; Attit.2 = Attitudes at T2; Inten.2 = Intentions at T2; Attit.3 = Attitudes at T3; Inten.3 = Intentions at T3.
* $p < .05$; ** $p < .01$; *** $p < .001$

Slightly different results of correlational analysis emerged in female drivers' group. Driving errors positively correlated with all variables of the study, except intentions to violate rules measured after training. Rule violations were found to be related with all variables across all measurements for women. Attitudes and intentions were found to be related to each other in all instances of measurements (Table 2).

Table 2. Correlations among driving behaviour, attitudes, and intentions in female drivers.

	Violations	Attit.1	Attit.2	Inten.2	Attit.3	Inten.3
Errors	.282**	.349***	.282**	−.108	.350***	.305**
Violations		.477***	.210*	.267**	.577***	.667***
Attit.1			.465***	.341**	.350***	.532***
Attit.2				.391***	.553***	.337***
Inten.2					.359***	.410***
Attit.3						.630***

Note: Attit.1 = Attitudes at T1; Attit.2 = Attitudes at T2; Inten.2 = Intentions at T2; Attit.3 = Attitudes at T3; Inten.3 = Intentions at T3.
* $p < .05$; ** $p < .01$; *** $p < .001$

To summarize the results of correlational analysis, violations while driving rather than errors were better explained by attitudinal variables for both males and females. Driving errors correlated with attitudes and intentions only for female drivers. The higher correlation coefficients were observed among variables of the same measurement (e.g. attitudes, intentions, driving errors, and violations after some independent driving).

3.2 The Role of Safety Attitudes and Intentions at Different Measurement Periods in Predicting Risky Driving of Novice Drivers

Correlational analysis revealed significant relationships among attitudes, intentions, and driving behaviour for both male and female drivers. But this did not allowed concluding that later behaviour might be predicted by previous attitudinal variables. As it was found that attitudes and intentions are interrelated across some measurement, this confound association might account for significant correlations among previous attitudinal characteristics and later behavioural variables. To make valid conclusion about possibility to predict later risky driving, it was necessary to control for interrelations among attitudes and intentions measured at the different instances. Several linear regression analyses were run separately for male and female drivers. Driving errors and violations were chosen as dependent variables. They were treated as distributed according to normal distribution (skewness and kurtosis did not exceed ±2, logarithmic transformation was applied to variable of violations to meet the condition). Attitudes at Times 1–3 and intentions at Times 2–3 were considered as independent variables.

Results of linear regression to predict driving errors for males and females were presented in Table 3. Regression model for driving errors in males was found to be statistically significant ($F = .194$; $df = 5$; $p = .013$). Independent variables together could explain 19.4% of variance of driving errors. After controlling for interrelations among independent variables only intentions to drive risky measured at the same time as self-reported errors was significant predictor. In other words, self-reported driving errors in male drivers' group could not be predicted in advance prior independent driving. For female drivers regression model was statistically significant (F = 7.451; df = 5; p < .001) and it allowed explaining 25.1% of driving errors' variance, none of independent variables were found to be predictors. There was only statistical tendency that risky attitudes measured before training could be related to driving error (p = .057). Due to low statistical power this result should be treated cautiously.

Table 3. The role of safety attitudes and intentions to predict driving errors in male and female groups.

	Males			Females		
	Stand. Beta	t	p	Stand. Beta	t	p
Attit.1	.024	−.175	.861	*.216*	*1.925*	*.057*
Attit.2	−.018	−.118	.907	.095	.920	.359
Inten.2	−.176	−1.270	.209	−.132	−1.379	.171
Attit.3	.056	.454	.652	.181	1.367	.174
Inten.3	**.378**	**2.932**	**.005**	.177	1.557	.122

Note: Attit.1 = Attitudes at T1; Attit.2 = Attitudes at T2; Inten.2 = Intentions at T2; Attit.3 = Attitudes at T3; Inten.3 = Intentions at T3.

Table 4. The role of safety attitudes and intentions to predict violations while driving in male and female groups.

	Males			Females		
	Stand. Beta	t	p	Stand. Beta	t	p
Attit.1	.134	1.083	.283	.093	1.027	.307
Attit.2	.027	.194	.847	*−.155*	*−1.858*	*.066*
Inten.2	.089	.709	.481	−.005	−.067	.947
Attit.3	**.340**	**3.070**	**.003**	**.304**	**2.845**	**.005**
Inten.3	**.273**	**2.342**	**.022**	**.480**	**5.228**	**<.001**

Note: Attit.1 = Attitudes at T1; Attit.2 = Attitudes at T2; Inten.2 = Intentions at T2; Attit.3 = Attitudes at T3; Inten.3 = Intentions at T3.

Violations as well as driving errors could be only explained by attitudes and intentions measured at the same time as behaviour for both male and female novice drivers (Table 4). Regression model for driving violations in male participants was statistically significant (F = 7.451; df = 5; p < .001) and parameters could predict 34.2% of variance of dependent variable. Thus, violations were associated with risky attitudes and intentions at Time 3 but could not predicted by intentions and attitudes assessed earlier. The model for female driving violation was significant as well (F = 23.245; df = 5; p < .001). Risky attitudes towards behaviour on the road and intentions to violate could explain 51.1% of variance of violations in female drivers' group. The statistical tendency that non-risky attitudes just after training could predict violations was observed.

4 Discussion

This study was aimed to predict risky driving behaviour of novice Lithuanian drivers during the first nine months of their independent driving by safety attitudes and intentions assessed before they were licenced to drive. Correlational analysis revealed significant relationships among later driving behaviour and risky attitudes as well as intentions, measured before driving training course, after driving training course and in follow-up for female drivers. It was found that driving errors were only related to intentions to take risk while driving in nine-month period after licencing among male novice drivers. Self-reported intentional violations correlated to risky attitudes measured before driving training course and in follow-up, as well as to intentions at follow-up. Similar results were found in previous studies [18–20, 25]. Despite significant relationships between attitudinal variables assessed before driving training and later driving behaviour, the conclusion that this behaviour could be predicted before licensing might be premature. Therefore, multivariate statistics was applied to consider covariance among attitudinal characteristics measured at different time points and ensure valid conclusions.

Results revealed that risky attitudes and intentions to risk on the road measured either before or after driving training course failed to predict self-reported risky driving during the first nine months after licencing when data of all measurements were considered. Quite similar results were found in previous research. Driving errors and speeding were not predicted by pre-driving attitudes in several novice drivers' samples [12, 25]. It could be assumed that to some extend prior licensing people have unrealistic expectations related to own abilities to drive and obey traffic rules which change when drivers start to drive independently and have to make own decisions on the road [28]. Results of this study pointed that either attitudes or intentions to behave in a certain way were not highly reliable indicators of future behaviour when people have no experience performing it in the context of driving. Therefore, they would not serve as characteristic which allows identifying at risk group for additional intervention or training before driving.

Nevertheless, driving behaviour in this study operationalized as self-reported driving errors and intentional violations had some relationship with risky attitudes and intentions even when more sophisticated statistics was applied. It was found that driving errors were explained by intentions to risk on the road measured at the same time point as the behaviour for males but not for females. When controlled for other attitudinal variables across measurements both attitudes and intentions evaluated after some driving experience remained significant indicators of intentional violations in novice drivers. The data might suggest that people possess attitudes and intentions that are congruent to their behaviour [13, 16, 17, 21]. It should be emphasized that intentional violations were better explained by attitudes and intentions than driving errors. This is in line with the idea that errors and violations are of the different psychological nature: errors might be associated with poorer cognitive abilities, like attentional inaccuracy, while violations might happen due to deliberate decision to behave in a risky way [34].

To sum-up risky driving could be better explained by attitudes and intentions, when drivers have already gained some driving experience. It might be presumed that driving experience shapes risky attitudes and intentions [27], not vice versa as it was expected by theory. The results imply the necessity to monitor what attitudes novice drivers gain during first year of independent driving. Some additional training during the first year of driving that emphasize analysis of emerging driving experience and own motives to obey traffic rules could be proposed.

References

1. Glendon, A.I., McNally, B., Jarvis, A., Chalmers, S.L., Salisbury, R.L.: Evaluating a novice driver and pre–driver road safety intervention. Accid. Anal. Prev. **64**, 100–110 (2014)
2. World Health Organization Homepage. https://www.who.int/violence_injury_prevention/road_safety_status/2015/en/. Accessed 04 Jan 2019
3. CARE database Homepage. https://ec.europa.eu/transport/road_safety/specialist/statistics_en. Accessed 04 Jan 2019
4. Lithuanian Road Police Office Homepage. http://lkpt.policija.lrv.lt/lt/statistika/keliu-eismo-taisykliu-pazeidimu-statistika. Accessed 04 Jan 2019
5. Scott-Parker, B., Watson, B., King, M.J., Hyde, M.K.: "I drove after drinking alcohol" and other risky driving behaviors reported by young novice drivers. Accid. Anal. Prev. **70**, 65–73 (2014)
6. Mann, H.: Predicting young driver behaviour from pre–driver attitudes, intentions and road behaviour. Doctoral thesis. Heriot–Watt University Applied Psychology School of Life Sciences, 314 (2010)
7. Vlakveld, W.P.: Hazard anticipation of young novice drivers; assessing and enhancing the capabilities of young novice drivers to anticipate latent hazards in roadandtraffic situations. Dissertation thesis, SWOV-Dissertatiereeks. Stichting Wetenschappelijk Onderzoek Verkeersveiligheid SWOV, Leidschendam (2011)
8. Martín-delosReyes, L.M., Jiménez-Mejías, E., Martínez-Ruiz, V., Moreno-Roldán, E., Molina-Soberanes, D., Lardelli-Claret, P.: Efficacy of training with driving simulators in improving safety in young novice or learner drivers: a systematic review. Transp. Res. Part F **62**, 58–65 (2019)
9. Useche, S., Serge, A., Alonso, F.: Risky behaviours and stress indicators between novice and experienced drivers. Am. J. Appl. Psychol. **3**(1), 11–14 (2015)
10. Beirness, D.J.: The relationship between lifestyle factors and collisions involving young drivers. In: New to the Road: Reducing the Risks for Young Motorists International Symposium, p. 17, June 8–11. UCLA Brain Information Service/Brain Research Institute, Los Angeles (1996)
11. Endriulaitiene, A., Šeibokaitė, L., Žardeckaitė-Matulaitienė, K., Markšaitytė, R., Slavinskienė, J.: Attitudes towards risky driving and Dark Triad personality traits in a group of learner drivers. Transp. Res. Part F **56**, 362–370 (2018)
12. Rowe, R., Maughan, B., Gregory, A.M., Thalia, C.E.: The development of risky attitudes from pre-driving to fully-qualified driving. Inj. Prev. **19**(4), 244–249 (2013)
13. Ajzen, I.: The theory of planned behavior. Organ. Behav. Hum. Decis. Process **50**, 179–211 (1991)
14. Engström, J., Hollnagel, E.: A general conceptual framework for modelling behavioural effects of driver support functions. In: Modelling Driver Behaviour in Automotive Environments, pp. 61–84. Springer, London (2007)

15. Chung, Y.S., Wong, J.T.: Investigating driving styles and their connections to speeding and accident experience. J. East. Asia Soc. Transp. Stud. **8**, 1–15 (2010)
16. Forward, S.E.: The theory of planned behaviour: the role of descriptive norms and past behaviour in the prediction of drivers' intentions to violate. Transp. Res. Part F: Traffic Psychol. Behav. **12**(3), 198–207 (2009)
17. Poulter, D.R., Chapman, P., Bibby, P.A., Clarke, D.D., Crundall, D.: An application of the theory of planned behaviour to truck driving behaviour and compliance with regulations. Accid. Anal. Prev. **40**(6), 2058–2064 (2008)
18. Chan, D.C., Wu, A.M., Hung, P.W.: Invulnerability and the intention to drink and drive: an application of the theory of planned behaviour. Accid. Anal. Prev. **42**(6), 1549–1555 (2010)
19. Moan, I.S.: Whether or not to ride with an intoxicated driver: predicting intentions using an extended version of the theory of planned behaviour. Transp. Res. Part F **20**, 193–205 (2013)
20. Rowe, R., Andrews, E., Harris, P.R., Armitage, C.J., McKenna, F.P., Norman, P.: Identifying beliefs underlying pre-drivers' intentions to take risks: an application of the theory of planned behaviour. Accid. Anal. Prev. **89**, 49–56 (2016)
21. Satiennam, W., Satiennam, T., Triyabutra, T., Rujopakarn, W.: Red light running by young motorcyclists: factors and beliefs influencing intentions and behaviour. Transp. Res. Part F **55**, 234–245 (2018)
22. Webb, T.L., Sheeran, P.: Does changing behavioural intentions engender behaviour change? A meta-analysis of the experimental evidence. Psychol. Bull. **132**(2), 249–268 (2006)
23. Ivers, R., Senserrick, T., Boufous, S., Stevenson, M., Chen, H.Y., Woodward, M., Norton, R.: Novice drivers' risky driving behaviour, risk perception, and crash risk: findings from the drive study. Am. J. Public Health **99**(9), 1638–1644 (2009)
24. Paris, H., Van den Broucke, S.: Measuring cognitive determinants of speeding: an application of the theory of planned behaviour. Transp. Res. Part F **11**, 168–180 (2008)
25. Mann, H.N., Sullman, M.J.M.: Pre-driving attitudes and non-driving road user behaviours: does the past predict future driving behaviour? In: International Conference on Driver Behaviour and Training, vol. 3, pp. 65–73, Dublin, Ireland (2007)
26. Boccara, V., Delhomme, C., Vidal-Gomel, C., Rogalski, J.: Time course of driving-skill self-assessments during French driver training. Accid. Anal. Prev. **43**(1), 241–246 (2011)
27. Ram, T., Chand, K.: Effect of drivers' risk perception and perception of driving tasks on road safety attitude. Transp. Res. Part F **42**, 162–176 (2016)
28. Helman, S., Kinnear, N.A.D., McKenna, F.P., Allsop, R.E., Horswill, M.S.: Changes in self-reported driving intentions and attitudes while learning to drive in Great Britain. Accid. Anal. Prev. **59**, 425–431 (2013)
29. Parker, D., Reason, J.T., Manstead, A.S.R., Stradling, S.G.: Driving errors, driving violations and accident involvement. Ergonomics **38**, 1036–1048 (1995)
30. Stelmokienė, A., et al.: Lietuviškosios vairuotojų elgesio klausimyno versijos psichometrinių rodiklių analizė [Psychometric properties of driver behaviour questionnaire Lithuanian version]. Int. J. Psychol. Biopsychosoc. Approach **13**, 139–158 (2013)
31. Iversen, H., Rundmo, T.: Risk-taking attitudes and risky driving behaviour. Transp. Res. Part F **7**(3), 135–150 (2004)
32. Parker, D., Manstead, A.S.R., Stradling, S.G., Reason, J.T., Baxter, J.S.: Intention to commit driving violations: an application of the Theory of Planned Behavior. J. Appl. Psychol. **77** (1), 94–101 (1992)
33. Özkan, T., Lajunen, T.: Why are young men risky drivers? The effects of sex and gender-role on aggressive driving, traffic offences and accident involvement among young men and women Turkish drivers. Aggressive Behav. **31**(6), 547–558 (2005)
34. Lajunen, T., Özkan, T.: Self-report instruments and methods. In: Porter, B.E. (ed.) Handbook of Traffic Psychology, pp. 49–53. Academic Press, Cambridge (2011)

Analysis and Evaluation of Public Transport Safety in Vilnius

Iveta Stanevičiūtė[1] and Vytautas Grigonis[2(✉)]

[1] Department of Roads, Vilnius Gediminas Technical University, Saulėtekio al. 11, 10223 Vilnius, Lithuania
[2] Road Research Institute, Vilnius Gediminas Technical University, Linkmenų str. 28, 08217 Vilnius, Lithuania
vytautas.grigonis@vgtu.lt

Abstract. The largest cities in Europe are improving the public transport systems and primarily focuses on efficiency, safety attractiveness and number of other indicators, because such indicators have a direct impact on passenger satisfaction. The number of travels by public transport is increasing, and it will bring new challenges in the road safety. Existing public transport infrastructure in the biggest Lithuanian cities is no longer able to satisfy increasing passenger traffic volume, therefore traffic accidents involving public transport are also increasing. There is no detailed analysis of traffic accidents in Vilnius public transport subsystem, therefore all possibilities to improve safety in public transport are not exploited. The results of the study should help to identify the main causes of traffic safety accidents and prepare action plan to improve the situation in other Lithuanian cities.

Keywords: Road safety in public transport · Dedicated bus lines · Road safety measures

1 Introduction

The aim of the study is to carry out initial assessment of public transport (PT) safety in Vilnius city and propose modern safety and priority measures in the most critical points. The study focuses on Vilnius city as initial point of departure and identified solutions of repetitive critical points could be employed in other Lithuanian cities. Further analysis needs to elaborate such critical points and concentrates on safety measures ensuring a higher level of traffic safety.

2 Review Best Practice and Relevance

The implementation of a compact urban planning strategy requires higher density of population. The coherency between compact city and public transport is very peculiar – in parallel with the development of this strategy, the supply of public transport to be increased [1].

A. Varhelyi et al. (Eds.): VISZERO 2018, LNITI, pp. 140–145, 2020.
https://doi.org/10.1007/978-3-030-22375-5_16

The city of Singapore is one of the most progressive in the field of PT development and safety. Solutions to improve Singapore's public transport system include Rapid Transit Systems (BRT), buses, trains and taxis. The Land Transport Service of the country focuses on the development and improvement of the bus network by implementing various measures, such as: rush hour PT lanes, common 24-h PT lanes, priority for buses at intersections, the priority of the bus, from the bus stop in the bay, the implementation of the horizontal marking (see Fig. 1). Studies have shown that new horizontal marking solutions alongside 202 bus stops have reduced bus delays to 73% [2].

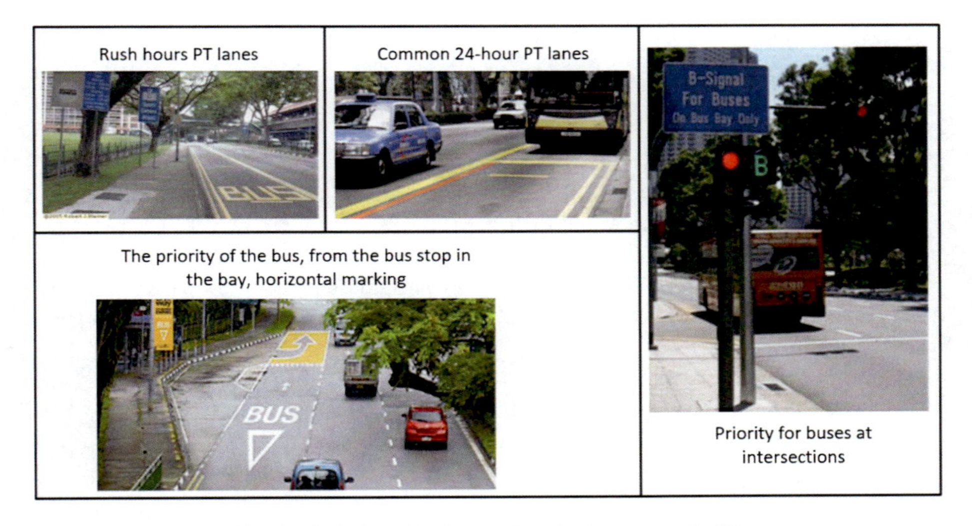

Fig. 1. Solutions for improving the bus network [3].

Some cities, such as Guadalajara, Bogota, Belo Horizonte, and Delhi, located in Mexico, Colombia, Brazil and India, have changed an existing dedicated public transport lanes or priority lanes in street center to BRT concept [4]. All the countries have achieved positive results on road safety over time. The introduction of the Guadalajara system in Bogotá resulted in a 46% reduction of traffic accidents and in a 60% reduction of fatal accidents. Although not all the countries were able to make a positive change at once. For instance, even after the installation of the BRT system on the main street in Belo Horizonte, the total number of accidents did not decrease. The number of accidents in the city of Delhi at the beginning of the installation is also increased. However, it should be noted, that it is always possible to improve the system and make it as safe as possible in different cities around the world by introducing additional traffic safety measures.

Starting from November 2012, the temporary Government of Lithuania approved the road traffic changes that allowed all drivers with ordinary vehicles carrying four and more passengers, drivers with electric vehicles use dedicated bus lines. The decision to validate this law has led to an increase in the purchase of electric vehicles in the country, but has caused more confusion in the streets, because more different types of PT dedicated lines has been introduced.

Vilnius City Municipality publish data on municipal transport indicators starting from 2016, such data describe the development of dedicated PT lanes in the city. First of all, the analysis of the data shows that dedicated PT lanes in the city of Vilnius were introduced in 2003. Secondly, the largest development took place in 2006 and 2013 and today there are 6 different types of PT lanes in the city. According to the inventory data of November 2015, there are 59 sections of various lengths of PT lanes in Vilnius streets [5]. The 5 sections (Zygimantų, Svitrigailos, Kalvariju and J. Tumo-Vaizganto streets) has a period of validity in order to ensure the effectiveness of the dedicated bus lanes, the period of validity has been set only from 7:00 a.m. to 20:00 p.m. in workdays.

3 Methods and Data Sources

Data analysis was accomplished with the help of ArcGIS. Data on traffic accidents before and after the installation of dedicated PT lanes on the streets (2013) were spatially analyzed. Data on road safety in Vilnius streets were collected from municipal enterprises "Vilniaus planas" and "Susisiekimo paslaugos", Police Department at the Ministry of Internal Affairs of the Republic of Lithuania and Road and Transport Research Institute.

Using the spatial analysis and assessment methodology, it is possible to determine the impact on road safety after the introduction of certain measures (see Fig. 2). This requires agglomerating data about traffic accidents on certain street 4 years before and 4

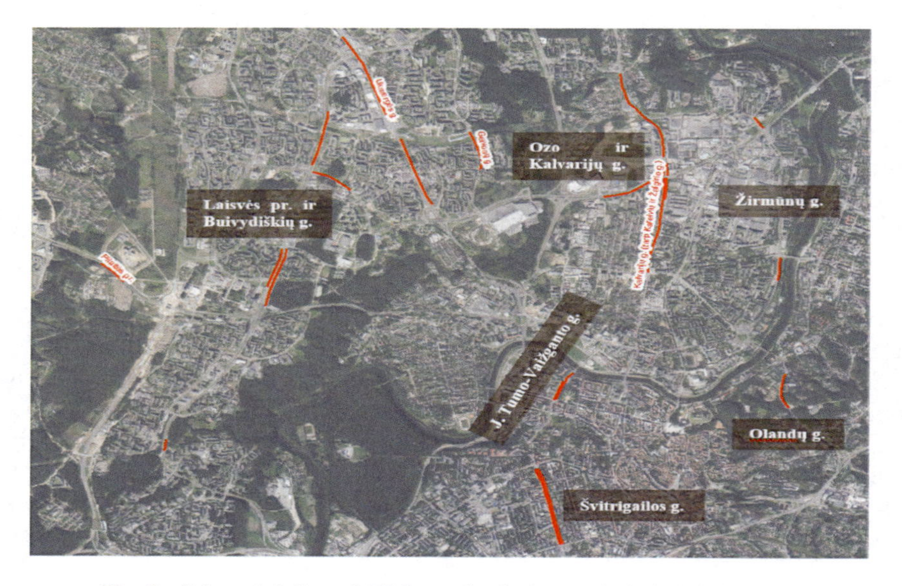

Fig. 2. Selected dedicated PT lanes for further analysis in Vilnius city.

years after the installation of the tool. The collected data (PT dedicated lanes installed in 2013) were evaluated using the statistical analysis method. In order to determine the changes in traffic safety caused by dedicated PT lanes, the data of traffic accidents in these street sections were analyzed in 2009–2013 period (before the introduction of PT lanes) and in 2013–2017 period (after the installation of dedicated PT lanes). The selected dedicated PT lanes are visualized in Vilnius city map.

4 Results of the Analysis and Further Research

The changes in the transportation systems of cities depends on the development of territories and the habits of people, i.e. commuting pattern. Therefore, the develop-ment of public transport and educational measures contribute to the creation of a sustainable environment in urban areas. When organizing the development of the public transport network, it is necessary to assess its impact on traffic safety [6] and to look for modern measures increasing the traffic safety in the city.

The analysis of accidents involving PT on selected dedicated lines has showed that the number of traffic accidents increased from 2009 to 2017. The Fig. 3 clearly shows the increase of traffic accidents involving PT.

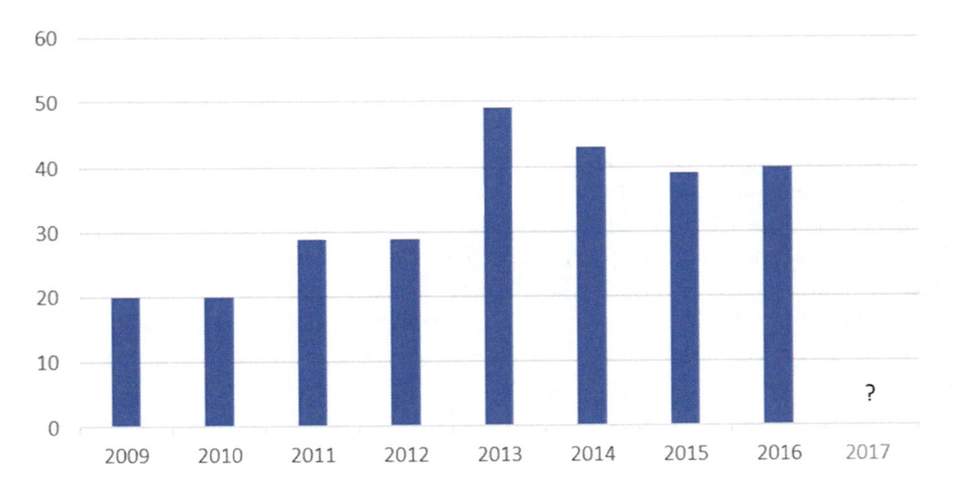

Fig. 3. Number of traffic accidents involving PT in Vilnius city from 2009 to 2017.

Although the traffic accidents data in 2017 are not yet available, the overall trend is significant, i.e. the number of traffic accidents increased by 33% when comparing years 2012 and 2016. Significant growth in traffic accidents occurred in 2013. When analyzing the change of traffic accidents, it can be seen that by 2013 traffic accidents increased every second year. After 2013, the number of injured and fatalities decreased by 2015, and a slight increase in traffic accidents has been observed since 2016.

The biggest changes in traffic accidents data occurred between 2012 and 2013. The biggest changes in traffic accidents after the introduction of dedicated bus lanes were observed at Ukmerge, Olandu and Kalvariju streets also in Laisvės av. When comparing traffic accident developments, it has been observed that the largest changes in traffic accidents occurred in the most overcrowded streets, although the introduction of PT lanes had to reduce the risk of conflict between various transport types (see Fig. 4).

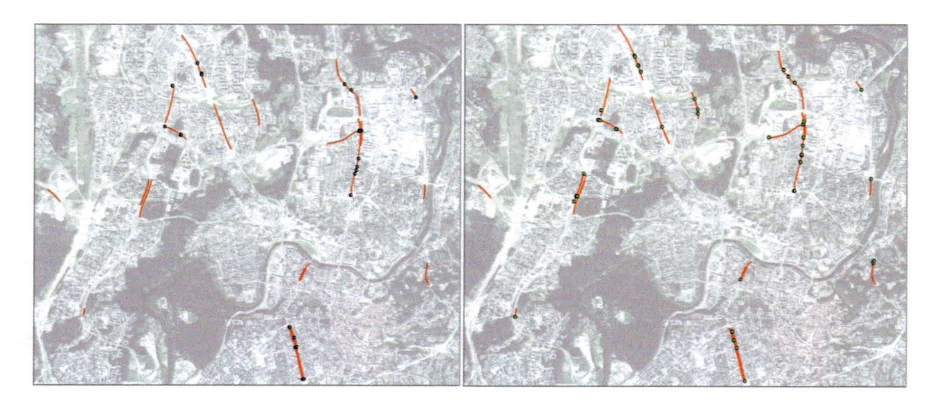

Fig. 4. Location of separate traffic accidents in 2012 (on the left) and 2013 (on the right) are marked with circles.

The number of PT traffic safety problems were identified in Vilnius streets such as Ukmerge, Olandu and Kalvariju, also in Laisvės avenue. Further analysis of modern safety measures such as rush hour PT lanes, common 24-h PT lanes, priority for buses at intersections, the priority of the bus, from the bus stop in the bay, the implementation of the horizontal marking will be performed on the basis of accident data analysis.

After the initial analysis and answering the main question - whether PT lanes have a negative impact on road safety – we can establish a further study plan including multi-criteria analysis [7]. This plan should include public transport systems of other Lithuanian cities also interurban buses. The obtained research results would help to formulate not only efficient, but also the right solutions from the point of view of traffic safety.

5 Conclusions and Recommendations

The methodology for collection of data for road safety impact assessment for PT is suggested and could be adopted in further research of biggest Lithuanian cities.

After analyzing the traffic accident data of dedicated PT lanes installed in the streets of Vilnius in 2013, it is found that these lanes have a negative impact on road safety. The installation of bus lanes on the streets has negative impact on traffic safety and number of accidents is growing significantly by 33%, although such a huge leap in the city's overall road safety situation is not observed.

When analyzing implemented sections of dedicated PT lanes, it is noticed that modern safety measures are not applicable on the streets and PT lanes are left isolated from common urban street environment. Moreover, there are no common vision on the development of dedicated PT lane core network and its safety issues.

The further hypothesis was developed that this situation affects the traffic incident growth trend in dedicated PT lanes sections and further work is related to the formulation of the concept of a unified PT lanes concept, which integrates dedicated PT lanes in common and safe urban street's pattern.

References

1. Clercq, F., de Vries, J.: Public transport and the compact city. Transp. Res. Rec. J. Transp. Res. Board **1735**, 3–9 (2000)
2. Singapore, L.T.A. Land Transport Master Plan (2013)
3. Haque, M.M., Chin, H.C., Debnath, A.K.: Sustainable, safe, smart—three key elements of Singapore's evolving transport policies. Transp. Policy **27**, 20–31 (2013)
4. Duduta, N., Adriazola, C., Wass, C., Hidalgo, D., Lindau, L.: Traffic Safety on Bus Corridors, Guidelines for integrating pedestrian and traffic safety into the planning, design and operation of BRT busways and bus lanes. EMBARQ **3**, 2–22 (2012)
5. ME „Susisiekimo paslaugos" strategy for PT (2015–2025). http://www.vilniustransport.lt/uploads/docs/SISP_Strategija%202015-2025_patvirtinta.pdf. Accessed 17 May 2018
6. Laurinavičius, A., Grigonis, V., Ušpalytė-Vitkūnienė, R., Ratkevičiūtė, K., Čygaitė, L., Skrodenis, E., Bobrovaitė-Jurkonė, B.: Policy instruments for managing EU road safety targets: road safety impact assessment. Baltic J. Road Bridge Eng. **7**(1), 60–67 (2012)
7. Nosal, K., Solecka, K.: Application of AHP method for multi-criteria evaluation of variants of the integration of urban public transport. Transp. Res. Procedia **3**, 269–278 (2014)

Links of Distracted Driving with Demographic Indicators

Oleg Bogačionok[1(✉)] and Alfredas Rimkus[1,2]

[1] Vilnius College of Technologies and Design, Olandų Str. 16, 01100 Vilnius, Lithuania
o.bogacionok@vtdko.lt

[2] Vilnius Gediminas Technical University, Saulėtekio al. 11, 03224 Vilnius, Lithuania
alfredas.rimkus@vgtu.lt

Abstract. During the research, distracted driving of drivers was selected. The aim of the research was to analyse the links of distracted driving with demographic indicators. To this effect, the behaviour of 1,896 drivers was observed while they were waiting for the green light at signal-controlled intersections in various cities of Lithuania. The analysis of the research results has revealed that the time of the day and such demographic variables as the gender and place of residence have an impact on the relation of drivers with distracted driving. The research results are ambiguous and claim that every second driver observed during the research is engaged in extraneous activity while driving a car. The most commonly encountered extraneous activities are talking on and manipulation of the phone, communication with passengers, and smoking. In addition, other kinds of extraneous activities have been observed, i.e. eating/drinking, checking one's appearance in the mirror, searching for fallen objects, cleaning the cabin, dozing off, throwing of rubbish through the window, using a computer, and etc. In conclusion, statistically significant dependencies between the types of extraneous activities and the demographic indicators of drivers as well as the time of the day have been determined.

Keywords: Drivers · Safe traffic · Distracted driving

1 Introduction

1.1 Relevance

Road accidents are regarded as one of the biggest threats to public health at international level. Currently, various strategies are being employed widely both in the world and in Lithuania so as to reduce the number of deaths resulting from road accidents, in particular by focusing on risky behaviour and its consequences. For instance, training programmes intended for safe driving are being implemented (Code 95); additional obligatory training is being applied to prevent risky behaviour. Due to persistent efforts of the Lithuanian government and authorities to reduce the number of road accidents, the number of deaths per 1m inhabitants decreased from 86 to 66 during 2013–2016 [1]. In addition, the total mortality rate has been constantly falling over the recent several years.

A. Varhelyi et al. (Eds.): VISZERO 2018, LNITI, pp. 146–158, 2020.
https://doi.org/10.1007/978-3-030-22375-5_17

In all European regions, the average number of traffic accident deaths is smaller by 1.8 unit compared to the world average (17.4 deaths are recorded in the world compared to 9.3 deaths in Europe per 100,000 inhabitants). However, certain European countries such as Romania, Bulgaria, Croatia, and Lithuania exceed both the European and the world average [2]. According to the World Health Organisation and the European Commission, the improvement of road safety still remains a priority area taking into account the spread and growth of negative consequences caused by risky driving in different countries of the world.

According to the World Health Organisation every year more than 1.25m people worldwide die in road accidents. Up to 50m people suffer fatal injuries or subsequently suffer from a disability. Injuries in road accidents have already become the main cause of death among 15–29 year-olds. Every third injured and every sixth killed pedestrian is hit at a pedestrian crossing. Road accidents cost approximately 3% of the gross domestic product [3] in the majority of states. Distracted driving was listed as one of the main causes of road accidents by the World Health Organisation [4]. The road accidents related to distracted driving represent around 22%, 1/3 of which are related to phone manipulation and communication with passengers [5, 6].

It is hard to detect whether a driver is engaged in these distracting activities; and also, with the wide range of these activities, the frequency of distracted driving is much higher than other dangerous driving behaviours, such as drinking and driving. According to the 2012 Traffic Safety Culture Index conducted by the AAA Foundation for Traffic Safety [5] over 67% of Americans indicated that distracted driving has become a greater problem today compared to 3 years ago, and also ranked distracted driving high in the list of safety concerns including aggressive drivers, drinking and driving, and others.

The literature analysis revealed that psychological driving aspects are widely researched by foreign authors and that such authors mostly concentrate on the personality qualities of drivers associated with risky driving as well as on psychological factors encouraging such risky driving [7, 8].

The object of this research is distracted driving. We have taken into consideration the fact that the sources confirm the impact of the demographic variables on distracted driving. The results of the research in question are ambiguous.

Distracted driving is defined as any action that can divert a driver's attention from driving; such actions can significantly increase the risk of driver's mistakes and accidents. The actions that increase distractedness while driving include eating, communicating with passengers, adjusting the radio, using mobile phones for making calls and sending text messages [9].

For example, from 1999 to 2008, the number of deaths resulting from distracted driving in the United States of America increased from 10.9% to 15.8%; a dramatic rise in the number of such deaths can be observed in 2005–2008 (up to 28.4%) [10]. In 2011, around 11% of deaths in the United States of America (up to 20 years of age) resulted from distracted driving, 21% of which was caused by the use of mobile phones [11]. However, the biggest problem is not speaking on the phone or typing text messages. The biggest problem is drivers surfing the net. Road users do this both when stuck in traffic jams and driving on motorways – mobile phones have become part of our lives. Therefore, road safety is now facing additional dangers. The laws forbidding the holding of mobile phones in hands while driving exist in 139 countries, whereas 31 countries prohibited both the ordinary use of mobile phones and their use with hands-free equipment.

Clause 20 of the Road Traffic Rules of the Republic of Lithuania indicates that, "Drivers of motor vehicles, tractors and self-propelled vehicles must not use mobile telephones if they are held by hand unless the engine of a standing vehicle is switched off". Which means that speaking on mobile phone, searching the internet, sending text messages or using hand held mobile devices in any other way is only possible if a vehicle is parked safely in a permissible location with its engine switched off. The aforementioned clause also indicates that, "Drivers of motor vehicles, tractors and self-propelled vehicles must avoid any actions not related with the driving of the vehicle". The Law on Road Traffic Safety of the Republic of Lithuania defines that driving of a vehicle means actions by which a person in a vehicle or on it controls a vehicle. Thus, neither eating, nor drinking, nor any other activities not related with the controlling of a vehicle can be performed.

The Code of Administrative Offences (CAO) of the Republic of Lithuania provides that the use of hand held mobile communication devices while driving results in a fine amounting from 60 to 90 EUR. Other activities causing distracted driving such as eating, smoking or communicating with passengers are not regulated by the CAO.

It is difficult to provide an explicit answer to the question as to how many traffic accidents in the Republic of Lithuania have been caused by the actions of drivers not related to driving – practically no one wants to admit that they hit another car because they were focusing on their ringing phone; or that they did not notice a pedestrian because when they were approaching a pedestrian crossing all their attention was directed at the ketchup on their hotdog; or that they drove into the opposite traffic lane because they wanted to check the news on a social network.

Today, when a traffic accident occurs, investigators do not determine whether a driver was using a phone at the time of the accident. In the future, investigators will do this and they will also consider the circumstances.

1.2 Review of Studies on Distracted Driving Behaviour

Because driving is already a multifaceted and complex task, mistakes and performance decrements such as reduced lateral control, failures to recognize and obey signage, reduced hazard response times, and inattentional blindness are easily introduced when attention is diverted away to complete secondary activities. This ultimately increases the probability that the distracted driver will be involved in a vehicle crash [12].

The research on distracted driving classifies the distraction factors into external ones, i.e. when attention shifts beyond the boundaries of a vehicle (e.g. billboards or pedestrians), also referred to as rubbernecking [11] and internal ones, i.e. when attention shifts to a certain action inside a vehicle (e.g. manipulating smart devices, radio or dashboard).

According to Rupp et al. [12], internal distraction factors are described in more detail compared to the external ones partly due to the fact that they are related to rapid spread of the vehicles with portable electronics (GPS, multimedia, LCD monitors,

Bluetooth and internet interfaces), which, in turn, provoke more serious and more frequently occurring distraction in a vehicle.

On the one hand, the sources of distraction are naturally more related with external factors and more related with a bigger number of traffic accidents [13]. One of the explanations for such intensity of external distraction factors is based on the fact that the variety of stimuli outside a vehicle demands a more complex processing load of the visual information that can occupy the entire attention of a driver and stop him from identifying priority information [14]. These distraction factors can be the following: roadway, air conditions, billboards, pedestrians, cyclists and motorcyclists, traffic accidents, observance of and search for road signs [12, 15].

On the other hand, it is described widely that such factors of distracted driving as communication with passengers, the use of mobile phones and eating while driving were the most common reasons for undesirable consequences [15].

The literature describes in great detail the dangers of distracted driving [16]. In most cases, driving disturbances (letting go of the steering wheel and difficulties to keep a vehicle in the right lane) [10, 17, 18] as well as disturbances in the identification of dangerous situations (longer realisation, reaction and breaking time) [9, 19] are discussed. The majority of such investigations include descriptions of experiments with simulators.

Other sources of literature focus more on the impact of mobile devices on the efficiency and safety of driving [20–23]. It is stated that the use of mobile phones is closely related with such demographic indicators as gender [18] and age [9, 17, 21, 24, 25]. For instance, new women drivers are two times more likely to use electronic devices than men drivers [13], whereas men up to 45 years of age and married men are more likely to use mobile devices than others [26]; women drivers and elderly drivers are more likely to feel lost when getting a phone call during driving [13]; young women drivers tend to suffer from more serious injuries resulting from traffic accidents caused by distracted driving than inexperienced men drivers [17, 27].

Some conducted research also attempted to evaluate the differences in driving efficiency between experienced and inexperienced drivers during distracted driving. Less experienced and younger drivers were more likely to engage in secondary activities compared to more experienced and more mature drivers. Middle-aged and elderly drivers were also more likely to engage in secondary activities: around 39% of elderly people (on average 73 year-old) and 43% of middle-aged drivers (on average 54 year-old) experienced greater difficulties when driving due to external and internal factors [21, 28]. Thus, during the previously carried out research, it was determined that distracted driving (mainly the use of a phone) had an impact on driving efficiency and safety among drivers of all age groups.

It is claimed that the use of mobile phones is related with more specific personality and behaviour qualities [7, 23, 29]. The drivers, who admitted participating in the activities of distracted driving more often than other drivers, did not consider it to be very dangerous [12]. Moreover, such drivers were more likely to underestimate potential danger and less likely to foresee a traffic accident caused by the consequences of distracted driving [10]. Young drivers [23, 24] and drivers of higher income level [22] are more likely to engage in distracted driving. Men drivers claim that they are often forced to use mobile devices while driving because of specific working conditions (e.g. couriers and long distance drivers) [26, 30]. The received results suggest that men commit more subjectively assessed deliberate breaches as they consider risky driving and violations of traffic rules to be potentially more useful rather than harmful [31].

Previously conducted research stated that the problem of distracted driving could be more difficult to deal with compared to other kinds of risky driving, such as speeding and drink-driving, as certain drivers consider inattentive driving behaviour as "social and useful" [8, 23, 27].

If traffic laws on drink-driving, seat-belt wearing, speed limits, helmets, and child restraints are not enforced, they cannot bring about the expected reduction in road traffic fatalities and injuries related to specific behaviours. Thus, if traffic laws are not enforced or are perceived as not being enforced it is likely they will not be complied with and therefore will have very little chance of influencing behaviour [4].

Effective enforcement includes establishing, regularly updating, and enforcing laws at the national, municipal, and local levels that address the above mentioned risk factors. It includes also the definition of appropriate penalties [4].

The transformation of current approach of a safe driving culture towards distracted driving could be one of the measures to reduce the number of traffic accidents. Even though there is no common, tangible definition of a safe driving culture, there is an agreement that a road safety culture focuses not only on risky behaviour and its consequences, but also on the changing of social norms, values and beliefs. Previous research [26] recommended creating an exhaustive programme involving various aspects of society in order to develop a safer driving culture. Zero driver distraction as well as better understanding and awareness of safety are considered to be the key target to achieve a better driving culture.

The analysis of the literature revealed the need to research more thoroughly the current driving culture related to distracted driving globally as well as in Lithuania. According to the evaluations by various authors, there is a link among separate factors (e.g. social, economic and demographic characteristics of participants), experience and attitude towards distracted driving (usually towards the use of mobile phones). The results have shown that carelessness, experience and driver behaviour are closely related. Moreover, the analysis suggests that the characteristics of the participants (e.g. income) have a great impact on their behaviour related to the issues of distracted driving. The results of the present article can be useful when preparing the measures intended for target groups of drivers (in particular young, inexperienced drivers) in order to affect the culture of distracted driving.

In Lithuania, the problem of distracted driving became relevant two decades ago, when the surge of smart technologies contributed to its fast development; however, no scientific research involving the aforementioned aspects in Lithuania has been discovered. The aim of this research is to analyse the links of distracted driving with demographic indicators.

2 Research Object and Methods

The observation lasted from 06/02/2018 to 23/03/2018. In total, 1,896 drivers were observed. The number included 1,314 (69%) men and 582 (31%) women. The researched people were observed at signal-controlled intersections in three cities of Lithuania (Vilnius (86%), Panevėžys (5%) and Kėdainiai (5%)) during daylight time. All general data was collected, i.e. the sex of the respondents, the bodywork type and the year of production of a car. Driver observation time: before 12 o'clock 808 (43%); after 12 o'clock 1,088 (57%). Bodywork type: passenger car 1,771 (93%); lorry 34 (2%); bus and trolleybus 91 (5%). Year of production of cars: <5 years 142 (7%); 5–10 years 696 (37%); >10 years 1,058 (56%).

3 Results and Their Discussion

The IBM SPSS statistical package was used to organize and process the research data collected through natural observation. The Shapiro-Wilk test was employed to check if the data meets the conditions of normal distribution. The statistical data analysis was conducted: frequencies, averages, standard deviations were calculated. The statistical conclusions were drawn on the basis of nonparametric criteria (Mann-Whitney U, Kruskal – Wallis H).

The Test on the Normality of Sample Probability Distribution. The information on the probability distribution type was determined with the help of the normality criterion of the Kolmogorov-Smirnov test. The conformity of variable distributions to the normal ones was checked. All distributions were $p = 0.000 < \alpha = 0.05$. Based on the received results it can be claimed that no distribution conformed to a normal distribution; therefore, the methods of not normal distribution (the Mann–Whitney U test for two independent samples and the Kruskal-Wallis H test for independent samples) are going to be employed for future research. A chi-squared test was used to determine the dependence among the samples.

While comparing the driver occupation estimates between men and women, a nonparametric Mann–Whitney U test for two independent samples was employed (see Table 1).

Table 1. Comparison of driver occupation estimates between men and women.

Occupation of drivers while waiting for the green light	Drivers gender	*N*	Rank average	*Z*	*p*
Eat, drink	Men	1,314	941.58	−2.48	0.01*
	Women	582	964.12		
Smoke	Men	1,314	962.38	−4.46	0.00*
	Women	582	917.16		
Communicate with passengers	Men	1,314	962.21	−2.98	0.00*
	Women	582	917.56		
Check their appearance	Men	1,314	927.44	−8.44	0.00*
	Women	582	996.06		

$*p \leq 0.05$

The table indicates the number of men and women in each group, a rank average (nonparametric substitute for an arithmetic mean) as well as Z and P values indicating statistical significance. Statistically significant differences in the estimates between men drivers and women drivers have been observed in four types of occupation ($p < 0.05$), i.e. eating, drinking and snacking, smoking, communicating with passengers and checking the appearance. With reference to the ranks indicated in the table, it can be stated that, from the point of view of statistical significance, women are more likely to eat, drink and snack as well as check their appearance (with 95% statistical guarantee) than men. Also, from the point of view of statistical significance, men are more likely to smoke and communicate with passengers compared to women. Other statistically significant differences have not been detected ($p > 0.05$).

While comparing the driver occupation estimates among the drivers driving different bodywork cars, a nonparametric Kruskal-Wallis H test for independent samples was used (see Table 2).

Table 2. Comparison of driver occupation estimates among drivers driving different bodywork cars.

Occupation of drivers while waiting for the green light	Bodywork type	N	Rank average	H	p
Do not engage in extraneous activities	Lorry	34	1,009.53	9.43	0.01*
	Bus, Trolleybus	91	1,090.73		
	Passenger car	1,771	940.02		
Smoke	Lorry	34	1,041.91	8.55	0.01*
	Bus, Trolleybus	91	923.34		
	Passenger car	1,771	948.00		
Communicate with passengers	Lorry	34	841.00	6.76	0.03*
	Bus, Trolleybus	91	903.51		
	Passenger car	1,771	952.88		
Other	Lorry	34	922.00	5.94	0.05*
	Bus, Trolleybus	91	984.51		
	Passenger car	1,771	947.16		

*$p \leq 0.05$

Statistically reliable differences have been observed in four driver occupations ($p < 0.05$). From the point of view of statistical significance, the intensity of bus and trolleybus drivers not engaging in any secondary activities clearly stands out (1,090.73). Lorry drivers are more likely to smoke (1,040.91), whereas passenger car drivers tend to spend their time waiting for the green light at the intersections communicating with passengers (952.88) more often than drivers of other types of vehicles. Bus and trolleybus drivers are more likely to engage in other activities, not specified in this observation (984.51).

While comparing the driver occupation estimates among the drivers driving at different times, a nonparametric Mann–Whitney U test for two independent samples was employed. The results are presented in the table below (see Table 3).

Table 3. Comparison of driver occupation estimates among drivers driving at different times.

Occupation of drivers while waiting for the green light	Driver observation time	N	Rank average	Z	p
Communicate with passengers	Before 12 o'clock	808	959.93	−5.03	0.00*
	After 12 o'clock	1,088	940.01		
Eat, drink	Before 12 o'clock	808	929.60	−3.89	0.00*
	After 12 o'clock	1,088	962.54		

$*p \leq 0.05$

The significances evident from the table indicate that only the estimates of such occupations as eating, snacking and drinking, and communicating with passengers differ ($p < 0.05$). With reference to the ranks indicated in the table, it can be stated that, from the point of view of statistical significance, drivers are more likely to eat, drink and snack in the afternoon, and communicate with passengers before. Other statistically significant differences have not been detected ($p > 0.05$).

While comparing the driver occupation estimates among the drivers driving the cars produced in different years, a nonparametric Kruskal-Wallis H test for independent samples was used. The results are given in the table below (see Table 4).

Table 4. Comparison of driver occupation estimates among drivers driving the cars produced in different years.

Occupation of drivers while waiting for the green light	Year of production of cars	N	Rank average	H	p
Use a mobile phone	<5 years	142	988.11	7.20	0.03*
	5–10 years	696	968.84		
	>10 years	1,058	929.80		
Look for things in the cabin	<5 years	142	935.00	7.75	0.02*
	5–10 years	696	941.81		
	>10 years	1,058	954.71		
	5–10 years	696	951.97		
	>10 years	1,058	947.09		

$*p \leq 0.05$

Statistically reliable differences have been observed in two driver occupations. Drivers driving the newest cars are more likely to use their mobile phones at the intersections than drivers of older cars (988.11). Drivers of the oldest cars are more likely to look for fallen objects in the cabin at the intersections (954.71).

While comparing the driver occupation estimates among the drivers in Vilnius and other big cities, a nonparametric Mann–Whitney U test for two independent samples was employed. The results are presented in the table below (see Table 5).

Table 5. Comparison of extraneous activity estimates between drivers in Vilnius and those in other cities.

Occupation of drivers while waiting for the green light	Driver observation locations	*N*	Rank average	*Z*	*p*
Do not engage in extraneous activities	Vilnius	1,623	932.75	−3.55	0.00*
	Other cities	273	1,042.11		
Use a mobile phone	Vilnius	1,623	967.39	−5.74	0.00*
	Other cities	273	836.17		
Eat, drink	Vilnius	1,623	952.30	−2.21	0.03*
	Other cities	273	948.50		
Take a nap	Vilnius	1,623	947.25	−2.49	0.01*
	Other cities	273	955.92		
Other	Vilnius	1,623	941.28	−4.91	0.00*
	Other cities	273	991.45		

$^{*}p \leq 0.05$

The table includes the number of the drivers observed in Vilnius and other big cities, a rank average (a nonparametric substitute for an arithmetic mean) as well as Z and P values indicating statistical significance. The significances evident from the table indicate that people in other cities of Lithuania are less likely to engage in secondary activities than people in Vilnius. Also, drivers in other cities are more likely to take a nap or engage in other activities, not specified in the observation ($p < 0.05$).

With reference to the ranks indicated in the table, it can be stated that, from the point of view of statistical significance, drivers in Vilnius are more likely to use their mobile phones, eat, drink and snack than drivers in other cities. Other statistically significant differences have not been detected ($p > 0.05$).

During the research, it has been determined that every second driver is engaged in secondary activities while driving a car (45%) (see Fig. 1).

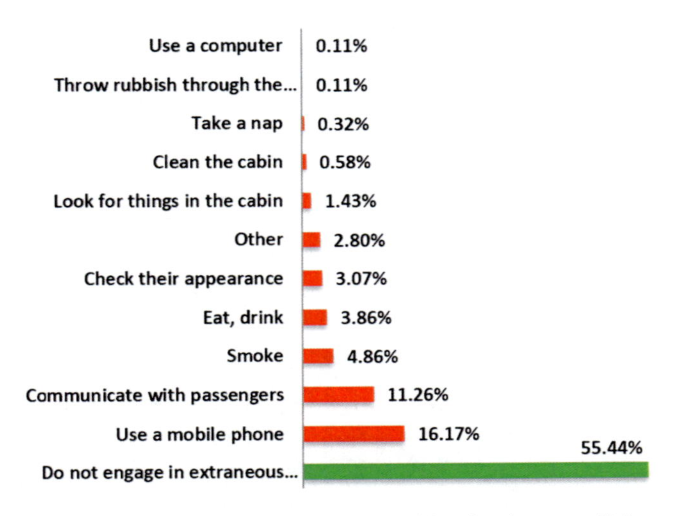

Fig. 1. Occupation of drivers while waiting for the green light.

The most common secondary activities include talking on and manipulating the phone (16.17%), communicating with passengers (11.26%), and smoking (4.86%). The following secondary activities have also been observed: eating/drinking (3.86%), checking one's appearance (3.07%), searching for fallen objects (3%), cleaning the cabin (0.58%), taking a nap (0.32%), throwing rubbish through the window (0.11%), using the computer (0.11%) and other not specified actions (2.8%). This confirms the afore-discussed literature results claiming that distracted driving is global and that it is a relevant issue for Lithuanian drivers.

The most recurrent secondary activities of men drivers and women drivers do not differ, i.e. the use of mobile phones (men – 36%, women – 37%). However, other results include statistically significant differences. For instance, if communicating with passengers is in the second (29%) and smoking – in the third place (14%) for men, checking their appearance is in the second (18%) and communicating with passengers – in the third place (18%) for women. This clearly contradicts other research results concerning the use of phones stating that men are more likely to use their mobile phones than women.

The research was conducted in three cities of Lithuania. Therefore, the secondary activities typical for drivers in these cities can be distinguished. The most popular secondary activities among drivers in Vilnius included the use of mobile phones (39%), communicating with passengers (24%) and smoking (11%). Drivers in Panevėžys were more likely to engage in communication with passengers (50%), the use of mobile phones (15%), smoking (9%) and other activities (15%). Finally, drivers in Kėdainiai also prefer communicating with passengers when driving (27%), using their mobile phones (10%), smoking (10%) and engaging in other secondary activities (27%).

There are more men drivers than women drivers in the research (men – 69%, women – 31%). The majority of the observed vehicles were passenger cars (93%). Also, buses (5%) and lorries (2%) driven mostly by men were observed.

The following car manufacturers are the most popular among drivers: Volkswagen – 16%, Toyota – 11%, Audi – 9%. BMW, Opel and Volvo collected 7% each; Ford, Mercedes-Benz and Škoda – 5% each; and finally, Citroen, Nissan, Peugeot and Renault – 3% each.

More than half of the observed cars were more than 10 years old (56%), the cars from 5 to 10 years old comprised 37%, and the cars up to 5 years old – 7%. Irrespective of the year of production of a car, the culture of drivers "violators" is basically the same, i.e. the use of mobile phones (33%, 40%, 45%) and communication with passengers (25%, 26%, 26%) remain in the first and the second places. However, if drivers of older cars are more likely to smoke (13%, 8%) and snack (8%), owners of the newest models are more likely to engage in checking their appearance (11%). Even though most of the cars observed during the research were more than 10 years old (56%), their drivers were more likely to drive more safely and 57% of them did not engage in any secondary activities. In contrast, owners of newer cars paid less attention to driving (54%) and more attention to secondary activities. Such results can indirectly support the observations of other authors stating that drivers with bigger income (i.e. owners of the newest cars in our research) are more likely to ignore the restrictions of secondary activities.

4 Conclusions

1. The observation results have revealed that practically every second (45%) driver observed during the research was engaged in extraneous activities while waiting for the green light at a signal-controlled intersection. The most commonly encountered extraneous activities are talking on and manipulation of the phone (36%), communication with passengers (25%), and smoking (11%).
2. The comparison of the indicators according to sex has shown that driving men smoke ($Z = -4.46$; $p = 0.00$) and communicate with passengers ($Z = -2.98$; $p = 0.00$) more often, whereas driving women check their appearance ($Z = -8.44$; $p = 0.00$) and have a drink or a snack ($Z = -2.48$; $p = 0.01$) more often. No statistically significant differences between the sexes have been observed in the indicators related to phone usage and other extraneous activities.
3. The comparison of the indicators among the drivers of different bodywork type cars have revealed that the bus drivers often abstained from extraneous activities ($H = 9.43$; $p = 0.01$), whereas the lorry drivers smoked more often ($H = 8.55$; $p = 0.01$), and the passenger car drivers communicated with passengers ($H = 6.76$; $p = 0.01$) and engaged in other extraneous activities ($H = 5.94$; $p = 0.05$) more often.
4. The comparison of the extraneous activity indicators among the drivers driving the cars produced in different years has shown that the drivers driving the newest (up to 5 year-old) cars use their mobile phones more often ($H = 7.20$; $p = 0.03$), whereas the drivers of older (10 year-old and more) cars rummaged around the cabin looking for items more often ($H = 7.75$; $p = 0.02$).
5. The drivers in the capital engaged in extraneous activities more often than those in peripheral cities ($Z = -3.55$; $p = 0.00$) and used their phones ($Z = -5.74$; $p = 0.00$) and snacked ($Z = -2.21$; $p = 0.03$) more often. The drivers in the province took a

nap ($Z = -2.49$; $p = 0.01$) and engaged in other extraneous activities ($Z = -4.91$; $p = 0.00$) more often while waiting for the green light.

6. The drivers driving before noon communicated with passengers ($Z = -5.03$; $p = 0.00$) more often, and snacked ($Z = -3.89$; $p = 0.00$) more often in the afternoon.

References

1. Lithuanian Road Administration under the Ministry of Transport and Communications. https://lakd.lrv.lt/lt/eismo-saugumas/eismo-ivykiu-statistika. Accessed 17 Sept 2018
2. Slavinskienė, J.: Rizikingai vairuoti motyvuojančių veiksnių ir asmenybės savybių sąsajos svarba paaiškinant vairuotojų, netekusių teisės vairuoti, rizikingą vairavimą. VDU, Kaunas (2018)
3. The Global status report on road safety. World Health Organization, 43–44 (2015)
4. Road traffic injuries. World Health Organization (2018). http://www.who.int/news-room/fact-sheets/detail/road-traffic-injuries. Accessed 01 Nov 2018
5. Distraction and Teen Crashes: Even Worse Than We Thought - AAA NewsRoom (2015)
6. Traffic safety facts: Young drivers. NHTSA. Washington (2015)
7. Bumgarner, D., Webb, J., Dula, C.: Forgiveness and adverse driving outcomes within the past five years: Driving anger, driving anger expression, and aggressive driving behaviors as mediators. Transp. Res. Part F **42**, 317–331 (2016)
8. Wilson, F., Stimpson, J., Tibbits, M.: The role of alcohol use on recent trends in distracted driving. Accid. Anal. Prev. **60**, 189–192 (2013)
9. Pope, C.N., Ross, L.A., Stavrinos, D.: Mechanisms behind distracted driving behavior: the role of age and executive function in the engagement of distracted driving. Accid. Anal. Prev. **98**, 123–129 (2017)
10. Carter, P., Bingham, R., Zakrajsek, J., Shope, J., Sayer, T.: Social norms and risk perception: predictors of distracted driving behavior among novice adolescent drivers. J. Adolesc. Health **54**, S32–S41 (2014)
11. Klauer, S.G., Guo, F., Simons-Morton, B.G., Ouimet, M.C., Lee, S.E., Dingus, T.A.: Distracted driving and risk of road crashes among novice and experienced drivers. N. Engl. J. Med. **370**(1), 54–59 (2014)
12. Rupp, M., Gentzler, M., Smither, J.: Driving under the influence of distraction: examining dissociations between risk perception and engagement in distracted driving. Accid. Anal. Prev. **97**, 220–230 (2016)
13. Foss, R., Goodwin, A.: Distracted driver behaviors and distracting conditions among adolescent drivers: findings from a naturalistic driving study. J. Adolesc. Health **54**, S50–S60 (2014)
14. Louie, J., Mouloua, M.: Predicting distracted driving: the role of individual differences in working memory. Appl. Ergon. **74**, 154–161 (2019)
15. Terzano, K.: Bicycling safety and distracted behavior in The Hague, the Netherlands. Accid. Anal. Prev. **57**, 87–90 (2013)
16. Wu, J., Xu, H.: The influence of road familiarity on distracted driving activities and driving operation using naturalistic driving study data. Transp. Res. Part F **52**, 75–85 (2018)
17. Bingham, C.R., Ehsani, J.P.: The relative odds of involvement in seven crash configurations by driver age and sex. J. Adolesc. Health **51**, 84–490 (2012)

18. Bingham, C.R., Zakrajsek, J., Almani, F., Shope, J., Sayer, T.: Do as I say, not as I do: distracted driving behavior of teens and their parents. J. Saf. Res. **55**, 21–29 (2015)
19. Stavrinos, D., Jones, J., Garner, A., Griffin, R., Franklin, C., Ball, D., Welburn, S., Ball, K., Sisiopiku, V., Fine, P.: Impact of distracted driving on safety and traffic flow. Accid. Anal. Prev. **61**, 63–70 (2013)
20. Choudhary, P., Velaga, N.: Analysis of vehicle-based lateral performance measures during distracted driving due to phone use. Transp. Res. Part F **44**, 120–133 (2017)
21. Engelberg, J., Hill, L., Rybar, J., Styer, T.: Distracted driving behaviors related to cell phone use among middle-aged adults. J. Transp. Health **2**, 434–440 (2015)
22. Nurullah, A., Thomasm, J., Vakilian, F.: The prevalence of cell phone use while driving in a Canadian province. Transp. Res. Part F **19**, 52–62 (2013)
23. Shaabana, K., Gaweeshb, S., Ahmedb, M.: Characteristics and mitigation strategies for cell phone use while driving among young drivers in Qatar. J. Transp. Health **8**, 6–14 (2018)
24. Gershon, P., et al.: Teens' distracted driving behavior: Prevalence and predictors. J. Saf. Res. **63**, 157–161 (2017)
25. Trivedi, N., Beck, K.: Do significant others influence college-aged students texting and driving behaviors? Examination of the mediational influence of proximal and distal social influence on distracted driving. Transp. Res. Part F **56**, 14–21 (2018)
26. Li, W., Gkritza, K., Albrecht, C.: The culture of distracted driving: evidence from a public opinion survey in Iowa. Transp. Res. Part F **26**, 337–347 (2014)
27. Merrikhpour, M., Donmez, B.: Designing feedback to mitigate teen distracted driving: a social norms approach. Accid. Anal. Prev. **104**, 185–194 (2017)
28. Parr, M., et al.: Differential impact of personality traits on distracted driving behaviors in teens and older adults. Accid. Anal. Prev. **92**, 107–112 (2016)
29. Braitman, K., Braitman, A.: Patterns of distracted driving behaviours among young adult drivers: exploring relationships with personality variables. Transp. Res. Part F **46**, 169–176 (2017)
30. Sinelnikov, S., Wells, B.: Distracted driving on the job: application of a modified stages of change model. Saf. Sci. **94**, 161–170 (2017)
31. Vieira, F.S., Larocca, A.P.C.: Drivers' speed profile at curves under distraction task. Transp. Res. Part F Traffic Psychol. Behav. **44**, 12–19 (2017). https://doi.org/10.1016/j.trf.2016.10.018. Accessed 01 Nov 2018

Investigation of Drivers' Comfort Factors Influencing Urban Traffic Safety

Artūras Kilikevičius[1](✉), Kristina Kilikevičienė[2], and Jonas Matijošius[3]

[1] Vilnius Gediminas Technical University, Institute of Mechanical Science, J. Basanavičiaus 28, 03224 Vilnius, Lithuania
arturas.kilikevicius@vgtu.lt

[2] Department of Mechanical and Material Engineering, Vilnius Gediminas Technical University, J. Basanavičiaus 28, 03224 Vilnius, Lithuania
kristina.kilikeviciene@vgtu.lt

[3] Department of Automobile Engineering, Vilnius Gediminas Technical University, J. Basanavičiaus 28, 03224 Vilnius, Lithuania
jonas.matijosius@vgtu.lt

Abstract. Vibrations and their impact significantly affect dynamic bus parameters. This article mainly focuses on determining the impact of internal and external factors exciting the bus system on comfort. Comfort is defined by amplitudes of vibrations of structural points and values of amplitudes manifesting in the frequency band. Experimental bus research – measurements of vibrations of significant structural bus points – was carried out in the article. Measurements of vibrations of significant structural bus points allow evaluating the dynamic parameters of low-floor buses and their impact on the comfort.

Keywords: Comfort factors · Bus vibrations · Statistical analyses

1 Introduction

Vibrations and their impact significantly affect dynamic bus parameters. Knowing how to avoid casualty done by vibrations and the resonance effect in the systems being developed is essential. Based on the conducted analysis of bus systems and their components and aiming to justify technical and technological solutions of newly developed city bus systems, the main focus of the article was to study dynamic properties of city bus systems and the reasons for their formation [1, 2].

The work mainly focuses on determining the impact of internal and external factors exciting the bus system on comfort. Comfort is defined by amplitudes of vibrations of structural points and values of amplitudes manifesting in the frequency band. Operation of the engine is considered an internal factor exciting the system, while excitations and a shock excitation occurring while driving – as an external factor [3, 4].

Vibrations of low-floor buses change in the course of their operation. Vehicle vibrations due to non-shocked vertical and rotational movement of the vehicle are transmitted to the vehicle driver and passengers [5, 6]. The work examines dynamic parameters of the main structural elements of the bus (suspension, frame, shock

A. Varhelyi et al. (Eds.): VISZERO 2018, LNITI, pp. 159–165, 2020.
https://doi.org/10.1007/978-3-030-22375-5_18

absorbers and airbags). During operation, inappropriately matched or chosen structural elements of buses lead to high vibration amplitudes, which are transmitted to the driver and the passengers. Each of these two elements is used to reduce vibration. Thus precisely matched elastic – dynamic properties of suspension elements – shock absorbers and airbags ensure lower vibration amplitudes [7, 8].

Considering the problems resulting from dynamic impacts on the bus design, the passenger and the driver described above, dynamic properties of low-floor buses must be assessed [3, 9].

The friendliness of city buses to the driver and the road during operation can be increased using the properly matched static and dynamic suspension component properties. Although soft suspension is desirable, limited distance between the body and the tire and changing riding height is also limited by the movement of suspension springs. Having matched the said parameters to the optimum, own system frequencies can be further reduced [10].

Normally, the term "riding quality" is used to describe vehicle vibrations in the vibration frequency range from 0 to 25 Hz. High frequency interferences are called "noise". Low-frequency vehicle vibrations are normally caused by uneven road surface. The vehicle itself, its transmission or engine vibrations for example, cause vibrations. However, such vibrations are of higher frequency and more related to noise rather than to riding comfort [11].

In addition to the impact on riding and passenger comfort, unwanted riding vibrations can also have a negative impact on health. Impact on health is often chronic and associated with prolonged exposure. Determining the quantity of these effects is more complex than determining their impact on the driver's comfort. There may be cases where improved comfort does not necessarily mean decreased negative impact on health, and vice versa. This further complicates the assessment of vehicle movement vibrations [12].

2 Research Methodology

Experimental methods form the basis of research. The research conducted in the article is based on the principles of theoretical mechanics, theory of vibrations and dynamics of movement. Key statistical calculations were made using the software package "Origin". The software package "Pulse" was used in carrying out experimental research and analysing the measurement results received. The package "Origin" was used for the analysis of measurement results received in experimental research.

Experimental material comprises data on characteristics of the system for measuring vibrations of the researched object and suspension suppression elements. Experimental research was carried out in JSC Vilniaus Viešasis Transportas (Vilnius Public Transport). Certified stationary and portable measuring equipment of the Danish firm "Brüel & Kjær" was used in the research. The equipment meets the requirements set by the standard of measurement of vibration parameters.

Experimental research of the design of low-floor buses. The aim of the research was to determine dynamic parameters of bus suspension and the body.

Experimental research of the design of low-floor city buses consisted of two parts: measurements of vibrations of significant points of low-floor city bus frame and suspension (under various ambient conditions: at various speeds, with engine turned off and on, and in presence of external excitation) and modal analysis of the body of low-floor city buses, the aim of which was to determine resonant frequencies of the system and to show modal forms.

Brüel & Kjær vibration measuring devices were used to measure dynamic parameters. Vibration measuring devices include: 1. DELL computer. 2. Portable Measurement Results Processing Equipment 3660-D. 3. Seismic accelerometer 8344 (frequency range 0.2–3000 Hz, sensitivity 2500 mV/g).

Accelerometers 8344 were attached to the appropriate place of the bus using magnets or guidance pads. The guidance pads were in turn attached to the respective bus place. Guidance pads allow rearranging accelerometers in the required direction solidly affixing them.

The received measurement signals were computer-processed using Origin 6 and Pulse software packages, calculating signal spectra, distributions and statistical parameters [13]:

Arithmetic average:

$$\bar{x} = \frac{1}{n}\sum_{i=1}^{n} x_i \tag{1}$$

Standard deviation:

$$S_X = \sqrt{\frac{1}{n-1}\sum_{i=1}^{n} (x_i - \bar{x})^2} \tag{2}$$

Standard average deviation (average square error of the arithmetic average):

$$S_{\bar{X}} = \frac{S_X}{\sqrt{n}} = \sqrt{\frac{1}{n(n-1)}\sum_{i=1}^{n} (x_i - \bar{x})^2} \tag{3}$$

Vibrations of significant points of the frame and suspension of low-floor buses were measured in order to determine dynamic parameters of the frame and suspension, which would reflect the properties of the bus as a system. Response of significant points of the mechanical structure of low-floor city buses to external factors, such as the impact of the engine's operation, was examined. In this research, the parts of the bus were assessed as solid and non-deformable, i.e. characteristics of the suppression system were evaluated. In order to be able to objectively assess the regularity of movement of bus parts, 6 frame and 4 suspension points were measured (Fig. 1) showing the regularity of movement of the front, the center and the end of the bus. When measuring, the regularity of movement of the front suspension was described by points 1P and 2P, while points 3P and 4P respectively described the regularity of movement of the rear suspension. The use of this number of measurement points allows evaluating both vertical and rotary suspension movements. Movements of the bus frame were evaluated at 6 points: 2 at the front, 2 at the centre and 2 at the back. The use of this number of measurement points allows evaluating both vertical and rotary frame movements. Frame vibrations were also examined due to the fact that the engine is affixed to the frame, which causes additional excitation during operation. Engine-induced excitation is the main source of noise resulting in resonant frequencies of higher forms and thus reducing comfort and safety.

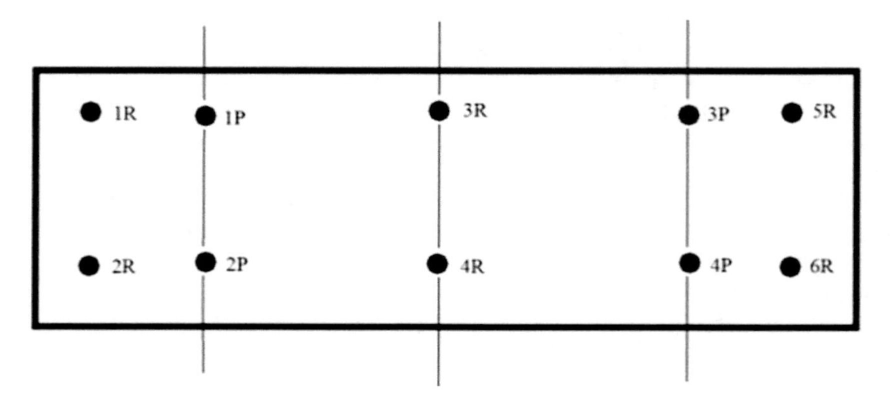

Fig. 1. The scheme of the arrangement of the measuring points in the bus frame and in the suspension (the bottom view of bus).

3 Research Results

Measurements of vibrations of the bus frame and suspension points (illustrated in Fig. 1) were conducted with engine turned on. The engine operated at idle speed, i.e. engine revolutions were 500 rpm. Measurements were made in vertical and horizontal directions. Results are illustrated in Fig. 2.

The measurement results (Fig. 2) show that having started the engine, vibrations up to 30 Hz in the vertical direction and up to 30 Hz in the horizontal direction manifest. Dominant acceleration amplitudes of the bus frame centre points 3R and 4R become apparent in the vertical direction at the frequencies of 9, 12, 23 and 24 Hz. Dominant acceleration amplitudes of rear suspension points 3P and 4P become apparent in the horizontal direction at frequencies of 2, 4, 8, 9, 13 Hz and 17–29 Hz.

Vibrations of the mechanical system of buses in running mode. In order to determine how vibrations are transmitted to bus structure while moving, the following measurements were made: the response of the mechanical system of the bus to road-induced excitation was measured. Many measurements were made: measurements with the bus moving at the speeds of 10 km/h and 30 km/h; with the bus suddenly braking and when starting the engine. The results of the experimental measurements show the characteristics of the vibrations transmitted by the bus system, i.e. frequencies of the acceleration amplitudes that are suppressed. 4 bus points were measured in the course

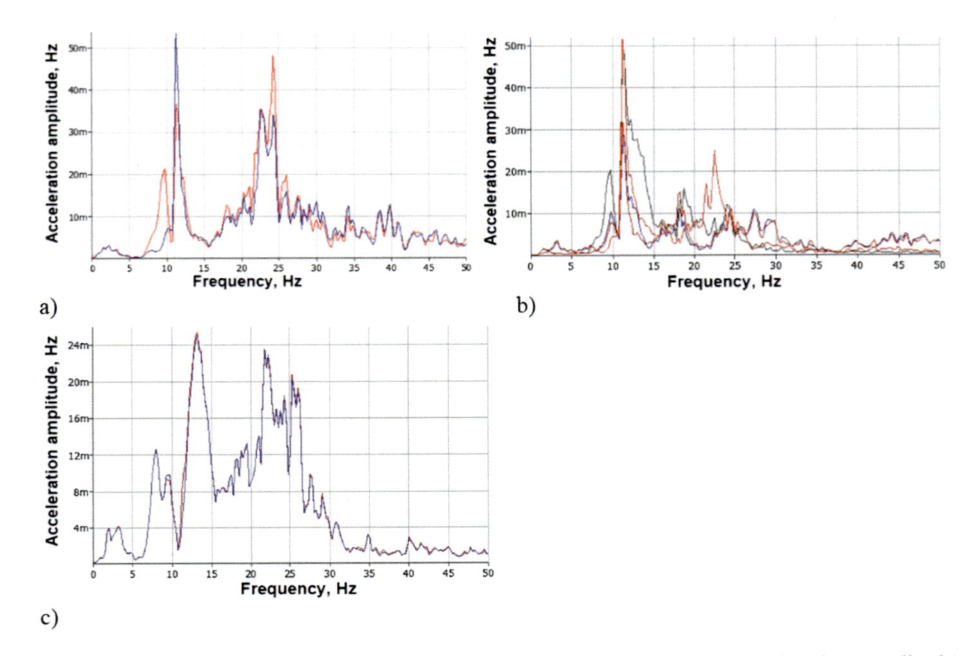

Fig. 2. The frame middle points spectral density of vibration vertical acceleration amplitude diagrams of: (a) points 3R and 4R; (b) points 1P, 2P, 3P and 4P; (c) points 3P and 4P when the engine is in operating mode

of the experimental research: the front suspension, the rear suspension, the center of the frame and the frame point above the front suspension. "Bröel & Kjær" accelerometers were used for measurements: 2 accelerometers 8341 (frequency range 0.3–3,000 Hz, sensitivity 100 mV/g) were used to measure suspension vibrations and 2 triaxial accelerometers 4506 (frequency range 0.6–3,000 Hz, sensitivity 100 mV/g) were used to measure frame vibrations (Fig. 2).

Figure 3 presents the measurement results received when the bus was moving at 10 km/h and 30 km/h, when it suddenly braked and at the time of starting the engine.

The assessment of the results presented in Fig. 3 reveals that with the bus moving at the speeds of 10 km/h and 30 km/h, dominant amplitudes of vertical vibrations manifested in the frequency range up to 12 Hz. In the assessment of vibrations of the front and rear suspension, dominant amplitudes of vertical vibrations of acceleration manifest in the frequency range from 6 to 12 Hz, and in the assessment of frame vibrations, they manifest in the range from 0 to 4 Hz, respectively. In order to assess vibrations of the bus design, examining vibrations in the range from 0 to 40 Hz would be expedient, because vibrations of higher frequency are caused by engine operation and they are assessed more like noise.

Vertical bus accelerations were determined experimentally under real conditions. The acceleration values are described by peaks, statistical parameters and the nature of changes in the acceleration signal in time.

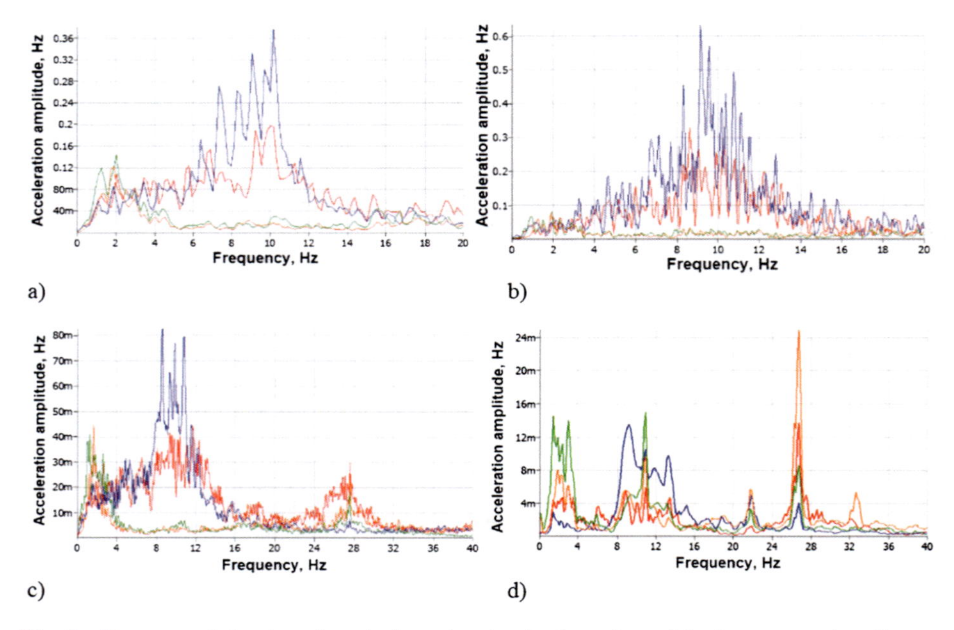

Fig. 3. The spectral density of vertical acceleration in the points of the bus suspension (front – red, rear – blue) and bus body (in center of bus body – green, over front suspension – orange) diagrams when: (a) bus rides speed 10 km/h; (b) bus rides speed 30 km/h; (c) bus rides and quick stops; (d) bus engine started up.

The assessment of the results revealed that the started engine (500 rpm) increases vertical vibration amplitudes by about 10 times (assessing front suspension vibrations) and 30 times (assessing rear suspension vibrations). The assessment of frame vibrations allowed determining that the started engine increases vertical vibration amplitudes about 10 times (assessing frame vibrations above the front suspension) and 30 times (assessing vibrations of the midpoint of the frame). These results indicate that the operation of the engine increases vertical vibration amplitudes of both the suspensions and the frame at the same ratio. Started engine causes high frequency vibrations in the range from 40 to 800 Hz.

Experimental measurements of vibrations of bus suspension and frame points made while in operation revealed that dominant acceleration vibration frequencies manifest up to 40 Hz. While moving (at 10 and 30 km/h), dominant frequencies of the suspension acceleration vibrations range from 6 to 12 Hz, while dominant frequencies of frame acceleration vibrations are up to 4 Hz.

4 Conclusions

The assessment of the experimental research results and their comparison with comfort requirements in public transport (these requirements determine the permissible frequencies and values of vibration amplitudes) revealed that when a bus is exposed to impact excitation and its engine is started, the acceleration values in the frequency spectrum are up to 0.045 m/s^2 (from 1 to 8 Hz) and up to 0.053 m/s^2 (from 8 to 25 Hz), and do not exceed the permissible values.

The evaluation of vibration parameters forming while moving allowed determining that the acceleration values in the frequency spectrum are up to 0.14 m/s^2 (from 1 to 4 Hz) and up to 0.04 m/s^2 (from 4 to 25 Hz), while the value of 0.14 m/s^2 shows that vibrations of such level lasting for more than 8 h significantly degrade the riding comfort.

References

1. Zhang, Y., Zhao, H., Lie, S.T.: A nonlinear multi-spring tire model for dynamic analysis of vehicle-bridge interaction system considering separation and road roughness. J. Sound Vib. **436**, 112–137 (2018)
2. Manes, A., Lumassi, D., Giudici, L., Giglio, M.: An experimental–numerical investigation on aluminum tubes subjected to ballistic impact with soft core 7.62 ball projectiles. Thin-Walled Struct. **73**, 68–80 (2013)
3. Ning, D., Sun, S., Du, H., Li, W., Li, W.: Control of a multiple-DOF vehicle seat suspension with roll and vertical vibration. J. Sound Vib. **435**, 170–191 (2018)
4. Kilikevičienė, K., Skeivalas, J., Kilikevičius, A., Pečeliūnas, R., Bureika, G.: The analysis of bus air spring condition influence upon the vibration signals at bus frame. Eksploat. Niezawodn. Maint. Reliab. **17**, 463–469 (2015)
5. Mohajer, N., Abdi, H., Nahavandi, S., Nelson, K.: Directional and sectional ride comfort estimation using an integrated human biomechanical-seat foam model. J. Sound Vib. **403**, 38–58 (2017)
6. Beijen, M.A., Heertjes, M.F., Butler, H., Steinbuch, M.: Disturbance feedforward control for active vibration isolation systems with internal isolator dynamics. J. Sound Vib. **436**, 220–235 (2018)
7. Rizvi, S.M.H., Abid, M., Khan, A.Q., Satti, S.G., Latif, J.: H ∞ control of 8 degrees of freedom vehicle active suspension system. J. King Saud Univ. Eng. Sci. **30**, 161–169 (2018)
8. Anastasopoulos, D., et al.: Identification of modal strains using sub-microstrain FBG data and a novel wavelength-shift detection algorithm. Mech. Syst. Signal Process **86**, 58–74 (2017)
9. Ataei, S., Miri, A.: Investigating dynamic amplification factor of railway masonry arch bridges through dynamic load tests. Constr. Build. Mater. **183**, 693–705 (2018)
10. Zhu, H., Yang, J., Zhang, Y.: Modeling and optimization for pneumatically pitch-interconnected suspensions of a vehicle. J. Sound Vib. **432**, 290–309 (2018)
11. Pollard, M.G.: Vehicle suspension systems and passenger comfort. In: Design for Passenger Transport, pp. 65–75. Elsevier (1979). https://doi.org/10.1016/b978-0-08-023735-0.50016-7
12. Abdelkareem, M.A.A., et al.: Vibration energy harvesting in automotive suspension system: a detailed review. Appl. Energy **229**, 672–699 (2018)
13. Devore, J.L., Farnum, N.R., Doi, J.: Applied statistics for engineers and scientists. Cengage Learning, Boston (2014)

Road Network Safety Ranking Using Accident Prediction Models

Vilma Jasiūnienė[1(✉)], Kornelija Ratkevičiūtė[1], and Harri Peltola[2]

[1] Vilnius Gediminas Technical University, Saulėtekio al. 11, 10223 Vilnius, Lithuania
{vilma.jasiuniene,kornelija.ratkeviciute}@vgtu.lt
[2] Technical Research Centre of Finland VTT, Vuorimiehentie 3, Espoo, Finland
Harri.Peltola@vtt.fi

Abstract. Road network safety ranking procedure and its implementation is not a strictly regulated road safety activity. This is one of the most flexible ways to determine the most effective and beneficious road safety investments. This paper analyses possibilities for implementing road network safety ranking according to the accidents prediction. The application of road network safety ranking procedure enables to prevent road accidents, i.e. to implement a proactive road safety activity. Road accident predictions are made using the empirical Bayes method, which is based on the assumption that in a similar environment with the similar traffic conditions the risk of accidents is similar. In order to implement this method, all roads of national significance of Lithuania were divided into homogeneous road sections and junctions. The homogenous road groups were determined based on 2012–2016 data of geometrical parameters of the road and traffic volume. Having estimated the predicted number of accidents for each homogenous road section, it is possible to calculate the predicted accident rate for each road. The authors of the paper, have predicted accident rate for the whole road, compiled a map of road safety levels for the trans-European roads in Lithuanian.

Keywords: Road safety · Road network safety ranking · Accident prediction model · Road infrastructure safety management

1 Introduction

Road accident may be caused by certain disturbances in one of the elements of safety system "Human – Vehicle – Road (environment)" or in the interaction between several of them. Accident statistics as well as the studies of many scientists (Elvik [2], Rolison et al. [11], Salmon et al. [12], Türker and Lajunen [13], Wierwille et al. [14], Jiménez-Mejías et al. [15]) show that the largest number of accidents occur due to the fault of road user, i.e. human factor. Understanding the fact that it is impossible to predict road user behavior, it is necessary to create such road and its environment, which would force the road user to choose the right solution in a dangerous traffic situation, the road user would have possibility to make minimum amount of wrong actions, and in the event of accident – its severity was the lowest possible. It takes long time to change the road user behavior and driving culture but by developing safer road infrastructure, it is

A. Varhelyi et al. (Eds.): VISZERO 2018, LNITI, pp. 166–176, 2020.
https://doi.org/10.1007/978-3-030-22375-5_19

possible to expect faster results in the field of road safety improvement. Besides, the adequate road infrastructure and traffic organization can not only protect road users from injuries but also form their behavior in a way to avoid road accident.

Seeking to pay a larger attention to one of the safety system elements "Road", in 2008 the European Parliament and the Council adopted the Directive 2008/96/EC on Road Infrastructure Safety Management [1]. In accordance with the provisions of the Directive each European Union (EU) Member State prepared and implemented in the country four procedures relating to road safety impact assessments, road safety audits, the management of road network safety and safety inspections. The procedures are divided into the already settled in the EU countries two groups of road safety activities – proactive and reactive. A reactive approach to road safety is associated with the identification of locations experiencing safety problems, problem definition, and implementation of road safety improvement measures. The aim of procedures of the proactive group is to detect and eliminate reasons, which may cause road accident. Both prevention and cure should be inherent elements of an overall road safety management system and ensure safety improvement during the whole service life of the road from planning to operation [5].

EU Directive is a legislative act the aims of which are obligatory to the Member States, though the forms and methods of their implementation can be chosen by the Member State itself. Based on this, implementation of the Directive 2008/96/EC procedures is not strictly regulated. This paper studies (analyses) possibility for the use of one of the procedures – network safety ranking and management.

2 Regulatory Framework for the Network Safety Ranking

According to the Directive 2008/96/EC, network safety ranking means a method for identifying, analyzing and classifying parts of the existing road network according to their potential for safety development and accident cost savings. Road Network safety ranking procedure should be performed periodically considering the 3–5 last years' data about traffic volume, accidents, road infrastructure changes etc. [8].

Analysis of various reports and legal documents of different EU Member States [7, 9, 10] shows that most frequently the network safety ranking is implemented by calculating certain accident indicators in reference population, i.e. in subsets of the road network, which have similar characteristics are expected to have similar safety performance [10] and by comparing (ranking) them between each other. The establishment of reference population is not strictly defined. Due to different calculation of accident indicators in road sections and junctions the road network is first of all divided into sections and junctions. When selecting criteria for a further division of sections and junctions into smaller groups, the main infrastructure parameters (road significance and road category, type of junction) and traffic parameters (annual average daily traffic, composition of traffic flow, etc.) are usually taken into consideration. Typical stages in road network safety ranking are: (1) collection of data on road parameters, traffic volumes, accidents and their severity; (2) definition of different road groups and junction groups; (3) dividing road network into homogenous road sections and junctions; (4) identification of hazardous road sections and performance of road network safety ranking; (5) in-office analysis of hazardous road sections and junctions and on-site observations of road user behavior.

In many cases, for the ranking of homogenous road sections the typical accident indicators are used, such as accident rate or accident density. In order to implement high accident concentration sections ranking in Lithuania [9] in each homogenous group the accident rate is calculated, which takes into consideration the severity of accident. Weight coefficients of accident severity are: 5 – accidents where at least one person was killed; 3 – accidents where at least one person was injured and undergoes in-patient treatment; 1 – accidents where people were injured and undergo (out-patient treatment).

Calculation of accident rate by taking into consideration the severity of accident is undoubtedly more advanced method compared to simple calculation of accident rate. However, this calculation defines the current network safety level. Based on the proactive approach that prevention is better than cure, the authors of this paper suggest to use accident prediction models to perform network safety ranking procedure. It would enable to rank the most hazardous road sections according to the predicted accident risk and, having implemented certain engineering traffic organization measures, to reduce this risk.

3 Network Safety Ranking Using Accident Prediction Models

Many scientists (Elvik [3], Hauer et al. [4]) point out that the empirical Bayes (EB) method is well-developed and widely used in the field of road safety. The EB method is based on the recognition that accident counts are not the only clue to the safety of an entity. Another clue is in what is known about the safety of similar entities i.e. in a similar environment with the prevailing similar traffic conditions the risk of accidents is similar. The literature analysis shows that the EB method should be a standard in identification of hazardous location [3, 6]. Using the EB method the expected number of accidents is determined by combining two information sources: (1) accident history on a specific road element (road section or junction), and (2) accident prediction model describing accident risk on the road elements similar in their environment. When using the EB method the expected number of accidents on a specific location is calculated by weighting the historical number of accidents on the location and the general expected number of accidents for similar road segments calculated by accident prediction models.

Vilnius Gediminas Technical University for already more than ten years has been co-operating with the Technical Research Center of Finland VTT in the field of accident analysis and accident prediction. Lithuania and Finland have been co-operating in developing tools for safety evaluations TARVA LT and ONHA LT. The purpose of the first common tool is to provide a method and database for the road network for (1) predicting the expected number of road accidents if no measures are taken to select locations for safety improvements, and (2) estimating the safety effects of road safety improvements in order to evaluate the cost-effectiveness of combination of road safety measures. This tool can be used also for the network safety ranking, i.e. to rank network safety by using not historic but the predicted number of accidents in road sections and junctions. The further sections describe the use of predicted accidents for the road network safety ranking.

3.1 Dividing Lithuanian Roads of National Significance into Homogenous Road Sections and Junctions

The scientific literature analysis shows that for the prediction purposes accident prediction model should be developed using 3–5 year's data. A larger amount of observation data allows to assess data dynamics and to make a more reliable prediction. Accident prediction model for Lithuanian road network has been developed based on 5-year data of observations.

The 2012–2016 data on technical road categories, road geometrical parameters, junctions, traffic volumes and historic road accidents was used. For the different type of road elements, the different mathematical prediction models were developed. Based on mentioned information the road network of Lithuania was classified into homogenous groups of road sections and junctions.

Description of the groups of road section. Road section is a part of the road between junctions. The length of road section depends on the road parameters which may influence occurrence of road accident. Homogenous road section groups were classified by the following criteria:

- road significance and road category;
- annual average daily traffic (AADT);
- composition of traffic flow.

Based on the above criteria 40 sub-groups of homogenous road section were determined. The roads of national significance of Lithuania (21,246 km) were divided into 10,505 homogenous road sections. Road groups for modelling were defined using data about technical road category, road significance and urban area, so that the road length in each group is reasonable high and groups will still reasonably well separate different kind of sections.

Roads differ from each other by their function, the level of traffic volumes, cross – section and geometric parameters. Technical categories of Lithuanian roads and their significance are described in legal document "Technical Road Regulation". The roads of national significance are classified to the main, national, and regional roads. According to the road technical category road network were divided into sub-groups: AM (highways), I – category roads (motorways), II, III and IV category roads (asphalt pavement) and V category roads (gravel roads). Data about the groups of homogenous road sections is given in Table 1.

Table 1. Data about the groups of homogenous road sections.

Road group	Number of road sections in road group	Length, km	Road sub-group	Number of road sections in road sub-group
AM	35	314	AM, Low AADT[(1)]	18
			AM, High AADT[(2)]	17
I	98	259	Roads with a median, Low AADT	76
			Roads with a median, High AADT	22
II–III	252	1,045	Main, Low AADT	183
			Main, High AADT	68
	472	1,414	National and regional, Low AADT	338
			National and regional, High AADT	134
IV, V and less	770	2,944	Regional, IV, Low AADT	528
			Regional, IV, High AADT	242
	601	1,199	Regional, IV, Low AADT	490
			Regional, IV, High AADT	111
	4,165	10,470	Regional, V and less, Low AADT	3,453
			Regional, V and less, High AADT	712
Urban roads	104	106	Main, Low AADT	85
			Main, High AADT	16
	820	759	National, Low AADT	674
			National, High AADT	141
	3,188	2,727	Regional, Low AADT	2,873
			Regional, High AADT	313
Total:	10,505[(3)]	21,236		10,494[(3)]

Notes: [(1)] – Low AADT – homogenous road sub-group where AADT is below the average of AADT in investigated road group; [(2)] – High AADT – homogenous road sub-group where AADT is above the average of AADT in investigated road group; [(3)] – missing data (AADT) in 11 road sections.

Description of the types of junctions. Homogenous junctions groups were classified by the following criteria:

- type of junction;
- road significance (in the case of three-leg (T) and four-leg (X) junctions);
- traffic flows at the junction (based on this criterion the junctions were grouped depending on the proportion of vehicles entering the junction from a minor road to the whole number of vehicles entering the junction).

Based on the above criteria, 2,784 junctions on the roads of national significance of Lithuania were divided into 21 sub-group of homogenous junctions (see Table 2).

Table 2. Data about the homogenous junctions groups.

Junction group	Number of homogenous junctions			Total
	0–5%[1]	6–15%	≥ 16%	
T, main roads[2]	99	47	24	170
T, national and regional roads	288	647	748	1,683
X, main roads	58	35	20	113
X, national and regional roads	52	193	374	619
Grade-separated	24	25	39	88
Roundabout	4	16	68	88
Regulated (traffic light)	3	6	14	23
Total:	528	969	1,287	2,784

Notes: [1] – proportion of incoming vehicles from other than the two main legs; [2] – the significance of the main road at the junction.

3.2 Accident Prediction Models for Road Sections and Junctions

Each homogenous group of road sections and junctions contains *n* of road sections/junctions. Comprehensive information gathered about each of them (accident history, road section length, traffic volume, etc.) enables to develop a mathematical accident prediction model.

Based on good modelling experience in Finland, new modelling type was tested for Lithuanian crossings and sections as well. Earlier, AADT groups were used as a classifying factor for creating the road groups that were assumed to have a constant accident rate. For example, rural 9 m wide main roads were earlier divided into three groups by AADT: (1) < 3,000 veh./day, (2) 3,000–6,000 veh./day and (3) more than 6,000 veh./day. To avoid irregularity around the group limits (3,000 veh./day and 6,000 veh./day) a continuous variable AADT was used in modelling. Respectively, the share of incoming vehicles from a minor road was used as a continuous variable for crossings, instead of using it as a grouping variable.

The type of the accident prediction model for the number of injury or fatal accidents on sections is:

$$\text{Accidents in five years (section)} = \text{Constant risk} \times \text{Effect of AADT on risk} \times \text{Vehicle kilometers during five years (expressed in 1,000 km)}.$$

The type of the accident prediction model for the number of injury accidents in junctions is:

$$\text{Accidents in five years (junction)} = \text{Constant risk} \times \text{Effect of the share of incoming vehicles from a minor road on risk} \times \text{Number of incoming vehicles during five years (expressed in thousands of arriving vehicles)}.$$

3.3 Ranking of Potentially Hazardous Road Sections

EU Directive on Road Infrastructure Safety Management [1] obliges Member States to carry out the network safety ranking of the trans-European road network. Republic of Lithuania is crossed by 6 trans-European roads (E67, E28, E77, E85, E262 and E272). The trans-European road network in Lithuania consists of 1,639 km of roads, i.e., almost all main roads of Lithuania belong to this network. Trans-European roads make 7.7% of all the roads of national significance of Lithuania.

Using EB method for all homogenous sections of trans-European roads, the number of which is 772, the expected accidents and the expected fatal accidents were calculated. Table 3 gives the relative predicted accident parameters, such as expected accident risk, expected accident density and fatal accident density, on trans-European roads that cross Lithuania.

Table 3. Accident prediction parameters in trans-European road network (in Lithuania).

Road No.	Accident parameters		
	Expected accident risk	Expected accident density	Expected fatality density
E67	4.34	11.50	2.98
E28	6.65	5.26	1.17
E77	10.30	4.31	0.28
E85	6.99	3.64	0.81
E262	8.33	14.00	1.98
E272	7.17	4.34	0.90
The average	7.30	7.18	1.35

Table 4. Most hazardous road sections in trans-European road network (in Lithuania).

Road No.	Road section, km		AADT, vpd	Homogenous road group[(1)]	Expected accident density, section	Expected fatality density, section
	From	To				
A6	13.62	14.41	25821	Urban, Main, High Heavy[(2)]	92.52	10.14
A5	0.00	0.42	45297	Main, II–III, High Heavy	33.41	9.55
A2	9.28	11.10	31398	I, High Heavy	26.32	8.77
A1	96.05	101.40	54024	I, High Heavy	54.73	7.66
A5	0.42	4.10	45297	I, High Heavy	39.39	7.61
A1	104.80	108.70	31344	AM, High Heavy	22.31	7.18
A5	89.37	95.01	9718	Main, II–III, High Heavy	21.97	6.91
A5	45.06	52.50	11715	Main, II–III, High Heavy	23.12	6.86
A5	57.66	58.10	15030	Main, II–III, High Heavy	20.45	6.82
A9	55.12	58.40	12514	Urban, Main, Low Heavy[(3)]	67.77	6.41
A8	8.33	23.61	7108	Main, II–III, High Heavy	19.96	6.28
A5	86.10	88.52	9718	Main, II–III, High Heavy	20.21	6.19
A16	93.01	95.12	8920	Main, II–III, High Heavy	19.00	6.18
A1	102.60	102.93	38400	I, High Heavy	27.44	6.10
A5	4.10	5.09	51167	I, High Heavy	37.22	6.04

Notes: [(1)] – see Table 1; [(2)] – High Heavy – homogenous road subgroup were heavy vehicle traffic volume is above the average of heavy vehicle traffic volume in investigated road group; [(3)] – Low Heavy – homogenous road subgroup were heavy vehicle traffic volume is below the average of heavy vehicle traffic volume in investigated road group.

The most hazardous trans-European road in Lithuania is E67 Via Baltica (274 km-long). It is the main trans-European motorway crossing Lithuania and running in the North-South direction. The traffic on this road is very intensive, almost one third of traffic is made of heavy vehicles. The cross section of the road does not meet existing traffic loads, since the major part of road has only two traffic lanes and only in the length of 40 km the opposite traffic is separated physically.

In order to carry out the network safety ranking in the trans-European road network the most hazardous high accident concentration sections were ranked. Table 4 gives 15 most hazardous road sections according to the expected fatality density, which is calculated based on the historical accident data of the section and using mathematical accident prediction models.

Analysis of the most hazardous trans-European road sections shows that 10 out of 15 sections are located on road E67 Via Baltica. This road for many years already is one of the most hazardous roads in Lithuania and is called "The Road of Death". However, in the nearest future the situation should change, since at present the road is under reconstruction where it is planned to change the cross section, to eliminate at grade intersections, to build 10 new viaducts and 2 roundabouts, to construct 3 grade-separated intersections, to build connecting roads for local transport and to install wild animal protection measures.

In order to demonstrate road safety levels on the main arteries of the country the map of road safety levels of the main roads was compiled (see Fig. 1).

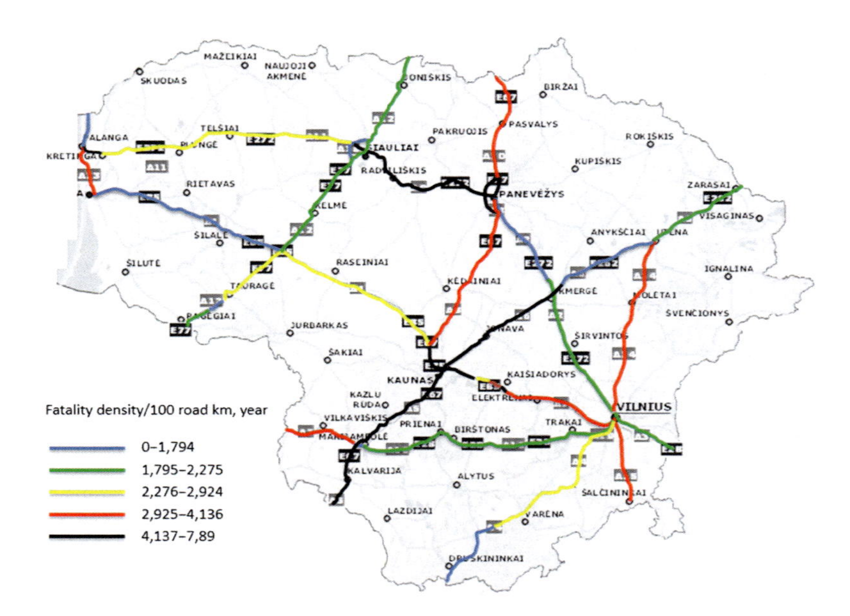

Fig. 1. Road safety levels of the main roads of Lithuania.

The main roads were divided into 5 road safety levels, which are based on the minimum and maximum fatality density values for each road: 1st level: 0–1.794 (total 315 km); 2nd level: 1.795–2.275 (total 357 km); 3rd level: 2.276–2.924 (total 383 km); 4th level: 2.925–4.136 (total 381 km); and 5th level: 4.137–7.89 (total 293 km). The intervals of road safety level values are based on the minimum and maximum expected fatality accident values and data sample.

Using the same methodology, the safety levels of the main roads were also determined in 2015. Comparing the safety levels of 2015 and 2018, there is a significant decrease in fatality density – 6.3 and 2.97, respectively. Network safety ranking and ranking of high accident concentration sections by using accident prediction models is a perfect tool for safety development and accident cost savings.

4 Conclusions

1. The road network safety ranking and ranking of high accident concentration sections is one of traffic safety tools (or procedures), which helps to set road safety priority interventions in case of limited financial resources. The road network safety ranking and ranking of high accident concentration sections by using accident prediction models give possibility to rank potentially hazardous road sections and to carry our accident prevention, and not only to solve the existing road safety problems. Implementation of the mentioned procedure could help to avoid the occurrence of new black spots as well as migration of the existing black spots.
2. The use of EB method for predicting accidents on the roads of national significance makes it possible to carry out road network safety ranking based on the expected number of accidents, and thus, to carry out accident prevention.
3. When predicting accident number in homogenous road sections and junctions the main parameters of road infrastructure and traffic were taken into consideration. Homogenous road sections groups were classified by the following criteria: road significance and road category, annual average daily traffic and composition of traffic flow. Homogenous junctions groups were classified by the following criteria: type of junction, road significance and traffic volume at the junction.
4. Ranking of high accident concentration sections on trans-European road network in Lithuania showed that 9 of 15 the most hazardous road sections belong to the E67 road Via Baltica. Fatality density on the main road A5, which is part of Via Baltica, is 2.3 times higher than the average fatality density in the whole main road network.
5. The map of road safety levels was compiled where the main roads were divided into five road safety levels based on minimum and maximum values of fatality density on each road.
6. A comparative analysis of the predicted accident indicators in the years 2007–2011 and 2012–2016 showed that the expected fatality density in the last 5-year period decreased from 6.3 to 2.97.

Acknowledgements. The discussed analysis tools were developed during several projects funded by the Lithuanian Road Administration under the Ministry of Transport and Communications.

References

1. Directive 2008/96/EC of the European Parliament and of the Council of 19 November 2008 on road infrastructure safety management, 9 p.
2. Elvik, R.: A framework for a critical assessment of the quality of epidemiological studies of driver health and accident risk. Accid. Anal. Prev. **43**(6), 2047–2052 (2011)
3. Elvik, R.: The predictive validity of empirical Bayes estimates of road safety. Accid. Anal. Prev. **40**(6), 1964–1969 (2008)
4. Hauer, E., Harwood, D.W., Council, F.M., Griffith, M.S.: Estimating safety by the empirical Bayes method. A Tutorial. Trans. Res. Record **1784**, 126–131 (2002)

5. Jasiūnienė, V., Čygas, D.: Road accident prediction model for the roads of national significance of Lithuania. Baltic J. Road Bridge Eng. **8**(1), 66–73 (2013)
6. Montella, A.: A comparative analysis of hotspot identification methods. Accid. Anal. Prev. **42**(2), 571–581 (2010)
7. Network Safety Ranking. TII Publications. Report GE-STY-01022 (2014). http://www.tiipublications.ie/library/GE-STY-01022-02.pdf
8. Ranking of High Accident Concentration Sections and Network Safety Ranking. BALTRIS WP 3.2.3 Development and dissemination of procedures for road infrastructure safety management. Report, p. 21 (2012)
9. Road network safety ranking. Legal document. Lithuanian Road Administration under the Ministry of Transport and Communications (2011)
10. Road safety manual. PIARC (2003)
11. Rolison, J.J., Regev, S., Moutari, S., Feeney, A.: What are the factors that contribute to road accidents? An assessment of law enforcement views, ordinary drivers' opinions, and road accident records. Accid. Anal. Prev. **115**, 11–24 (2018)
12. Salmon, P.M., Lenné, M.G., Stanton, N.A., Jenkins, D.P., Walker, G.H.: Managing error on the open road: the contribution of human error models and methods. Saf. Sci. **48**(10), 1225–1235 (2010)
13. Türker, Ö., Lajunen, T.: A new addition to DBQ: Positive driver behaviours scale. Transp. Res. Part F Traffic Psychol. Behav. **8**(4–5), 355–368 (2005)
14. Wierwille, W.W., Hanowski, R.J., Hankey, J.M., Kieliszewski, C.A., Lee, S.E., Medina, A., Keisler, A.S., Dingus, T.A.: Identification and evaluation of drives errors: Overview and recommendations. Report no. FHWA-RD-02-003. US Department of Transportation, Federal Highway Administration (2002)
15. Jiménez-Mejías, E., Martínez-Ruiz, V., Amezcua-Prieto, C., Olmedo-Requena, R., de Dios Luna-del-Castillo, J., Lardelli-Claret, P.: Pedestrian- and driver-related factors associated with the risk of causing collisions involving pedestrians in Spain. Accid. Anal. Prev. **92**, 211–218 (2018)

Vehicle Body Side-Slip Angle Evaluation and Comparison for Compact Class Vehicles

Robertas Pečeliūnas(✉) and Vidas Žuraulis

Vilnius Gediminas Technical University,
J. Basanavičiaus 28, 03224 Vilnius, Lithuania
{robertas.peceliunas,vidas.zuraulis}@vgtu.lt

Abstract. Compact class vehicles form a large share of transport in these days' European cities; therefore the stability of such kind of vehicles is an important issue of road traffic safety. In the article, the results of experimental research on compact class vehicle stability and road holding experimentally as well as computer modelling are evaluated. A double lane-change manoeuvre is selected for testing (ISO 3888-2). The information obtained from sensors was recorded by the "Corrsys-Datron DAS-3" data collecting device and then processed in common time scale upon applying "TurboLab" computer program. During the experiment, the vehicle's lateral acceleration, body oscillations and its rates were measured. The projections of longitudinal and lateral velocities were measured as well. According to the calculation method, the angles of each vehicle's side-slip were calculated. The vehicle's lateral acceleration, roll angle and roll rate objectively assess the comfort and freedom of movement of the sprung mass, but the stability of the vehicle is directly represented by the angle of the side slip.

Keywords: Vehicle dynamics · Compact class · Side-slip angle · Roll rate · Vehicle stability

1 Introduction

The road transport provides abundant advantages, as compared to other modes of transport; however, the increasingly growing number of motor vehicles causes a considerable danger. We suffer the biggest and mostly painful losses caused by day-to-day traffic events when people are injured or die. A safety of a vehicle is assessed according to various parameters, such as braking efficiency, lateral stability, riding qualities, active and passive protective means et cetera.

Manufacturers of vehicles pay a considerable attention to safe traffic and accident prevention: they permanently improve reliability, controllability and safety of vehicles, i.e. create a vehicle that's behaviour on driving should be predictable to a larger extent and enable avoiding emergency situations more easily. However, the practice shows traffic events to be an enormous problem yet, so an assessment of the interaction between the driver and the vehicle is a task of high importance. These two components mostly impact an appearance of conditions favourable for traffic events. In the presented research, stability and controllability of cars are assessed in an experimental way.

A. Varhelyi et al. (Eds.): VISZERO 2018, LNITI, pp. 177–187, 2020.
https://doi.org/10.1007/978-3-030-22375-5_20

In order to examine the influence of car type upon the casualty rate, car models have been divided into six types, ranging from 'Minis and Superminis' to '4 × 4 s and people carriers'. The types are broadly ordered by size, although the 'Sports cars' type is somewhat anomalous. The numbers of new cars of the smallest and largest types have grown in recent years. Exploratory analyses showed that the driver casualty rate falls markedly with size of car. In car–car collisions, the driver casualty rate also rises markedly with the size of the other car. Consequently, the relative risks faced by the drivers of two cars which collide are even more sensitive to the relative size of the two cars. Statistical models were fitted to identify the separate effects of car type on risk in car–car collisions. The results showed that:

- the mean risk of death for the driver of the smallest of the 6 types of car was 4 times higher than the risk for the largest type;
- the mean risk of death for a driver in collision with the largest type of car was over twice the risk when in collision with the smallest type [2].

In course of creation and improvement of stability systems for vehicles, computer models are developed; they are examined by real tests upon applying a double lane-change manoeuvre and other manoeuvres [11]. While evaluating the riding qualities of a car, a double lane-change manoeuvre was found to be the most appropriate method [13]. During the said manoeuvre, four driving motions (turning, straitening, returning and stabilization) are performed and analyzed. Because it is a manoeuvre of open-loop type, the skills of the driver are assessed in addition as well.

The methodology for assessing the yaw rate of a car by virtual sensors is described in Emirler et al. [6]. Active safety systems, such as electronic stability programme (ESP), are important and are usable for maintaining the riding stability of vehicles. The yaw rate is the key parameter for assessing the control of the car's yaw stability.

Minghui et al. [10] developed an algorithm for side-slip angle calculation. The said angle is highly important for stability of the vehicle. The authors state that a direct measurement of the side-slip angle of a vehicle requires high-priced equipment; however, this task may be accomplished simply and less costly upon applying simulation programmes.

For an accurate establishing the side-slip angle in simulation of the manoeuvre, Wang et al. [14] refused examining the actions of the driver in their research. Upon applying a modified model of "Dugoff" tire calculated on the base of a feedback, the double lane-change manoeuvre at the driving velocity of 26 m/s was unsuccessful.

Baffet et al. [1] were involved in a process of assessment of the slip-angles and the forces acting between the tires and the pavement. The process of assessment of the said forces is based on the series of two blocks: the task of the first block was an assessment of forces acting the tires and the road, whereas the second block "Kalman Filter" enables selecting the values of the slip-angle measured by the sensors.

While examining the lateral dynamics of cars, the research performed by Slaski [12] is notable. In the work, the double lane-change tests upon applying several damping levels according to the limitary characteristics of the shock-absorbers and several tests upon using an adjustable controller and intensive varying the level of damping according to some chosen signals defining the dynamic properties of the car were described.

In the work by Whitehead et al. [15], the rollover characteristics of vehicles were examined upon applying computer-based simulation. The accuracy of computer-based simulation was verified by its comparing with the experimental data. The performed tests enable establishing an ability of a vehicle to drive around an obstacle having appeared on the road with a certain velocity and to establish the lateral forces. In course of the research, dynamic simulation was proved to be an effective and accurate method than enables to analyze vehicles and assess their dynamics.

Comfort of a vehicle and driving stability depend on the stiffness of the suspension and the characteristics of the damping elements. In the paper by scientists Cronje and Els [5], the double lane-change manoeuvres were applied for examining the influence of a stabilizer on the driving properties of the vehicle while moving on road surface of varying quality. When active stabilizing elements were adapted for a land-rover, the side-slip angle reduced by 40–70% during the manoeuvre and was followed by improvement of the controllability and comfort.

An assessment of the side-slip angle is of a high importance for stability of the vehicle. The aim of this research is a calculation side-slip angle of vehicles according to the projections of the longitudinal and lateral velocities as well as the yaw rates fixed by the sensors during the experimental tests and upon applying the chosen methodology of calculation.

2 The Theoretical Model of a Vehicle

While analyzing the general stability of a vehicle, it is convenient to use a model that unites the rollover dynamics and horizontal dynamics of the vehicle (Fig. 1).

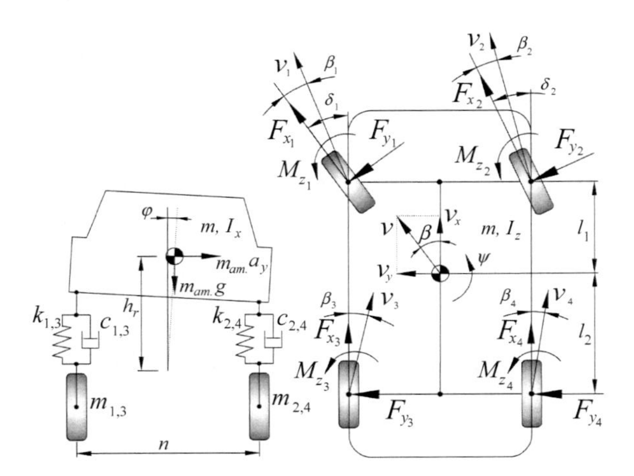

Fig. 1. Four degrees of freedom vehicle model.

In this four degrees of freedom model, lateral vibrations caused by the lateral force are described [9]:

$$F_y = F_{y_1} + F_{y_2} + F_{y_3} + F_{y_4} = m \cdot a_y; \tag{1}$$

where F_{y_i} – the lateral cohesion force appearing in the wheel contact; m – total vehicle mass; a_y – the acting lateral acceleration.

In the horizontal model of a vehicle, side-slip angles β_i appear upon action of lateral forces. The said angles show a discrepancy between the longitudinal plane of the relevant wheel and the real direction of movement defined by the velocity vector $\vec{v}_i$. The total side-slip angle of a vehicle body in the centre of gravity is marked as β.

Upon ignoring the loads caused by the longitudinal sliding of wheels and the loads caused by vehicle racing or braking, the equations of movement of this model according to the Newton's second law shall be written as follows:

$$m\left(\dot{v}_x - v_y\dot{\psi}\right) = F_{x_1}\cos\delta_1 - F_{y_1}\sin\delta_1 + F_{x_2}\cos\delta_2 - F_{y_2}\sin\delta_2 + F_{x_3} + F_{x_4}; \tag{2}$$

$$m\left(\dot{v}_y + v_x\dot{\psi}\right) = F_{x_1}\sin\delta_1 + F_{y_1}\cos\delta_1 + F_{x_2}\sin\delta_2 + F_{y_2}\cos\delta_2 + F_{y_3} + F_{y_4}; \tag{3}$$

$$\begin{aligned} I_z\ddot{\psi} = {} & (F_{x_1}\sin\delta_1 + F_{y_1}\cos\delta_1 + F_{x_2}\sin\delta_2 + F_{y_2}\cos\delta_2)\cdot l_1 - \left(F_{y_3} + F_{y_4}\right)\cdot l_2 - \\ & \left(F_{x_1}\cos\delta_1 - F_{y_1}\sin\delta_1\right)\cdot\tfrac{n}{2} + \left(F_{x_2}\cos\delta_2 - F_{y_2}\sin\delta_2\right)\cdot\tfrac{n}{2} + \sum M_{z_i}; \end{aligned} \tag{4}$$

$$I_x\ddot{\varphi} - I_z\ddot{\psi} = m_{am.}h_r\left(\dot{v}_x + v_y\dot{\psi}\right) + m_{am.}h_r g\varphi - \left(k_{pr.} + k_{gal.}\right)\varphi - \left(c_{pr.} + c_{gal.}\right)\dot{\varphi}; \tag{5}$$

here: m – total vehicle mass; v_x – longitudinal velocity component; v_y– lateral velocity component; ψ – yaw angle of sprung mass; F_{x_i} – longitudinal friction force; δ_i – wheels steering angle; $I_{x,z}$ – body roll inertia; $l_{1,2}$ – centre of gravity from the front and rear axle; n – track; M_{z_i} – tire aligning moment; φ – roll angle of sprung mass; h_r – distance from center of gravity to body roll axis; g – gravitational constant; $k_{pr.,gal.}$ – roll stiffness of the front and rear axle; $c_{pr.,gal.}$ – roll damping of the front and rear axle.

While driving on turning a tire loaded with a vertical force and a lateral force, its zone contacting with the road surface extends and the asymmetrically distributed reactions create the tire aligning moment M_{z_i}:

$$M_{z_i} = F_{y_i} \cdot t_{x_i}; \tag{6}$$

where t_{x_i} – the shift of the reaction of a deformed tire.

The tire aligning moment resists to the side-slipping.

3 The Computer-Based Simulation of Vehicle Movement Manoeuvre

Computer-based simulation is intended for establishing vehicle lateral acceleration, roll angle of sprung mass, side-slip on driving with velocities of 50, 60 and 70 km/h and for examining whether all the vehicles under testing are able to perform a double lane-change manoeuvre. The above-mentioned velocities had been chosen for singling out the cases (on a comparison of the obtained movement parameters) when movement occurs without skidding of the wheels and is affected by the side-slip angle only in the beginning of sliding or total sliding of the wheels.

A double lane-change manoeuvre (Fig. 6) is simulated in accordance with standard ISO 3888-2 on a dry asphalt pavement upon taking into account the height of irregularities of the road microprofile. The technical characteristics of the simulated vehicles are provided in Table 1 [3]. The lateral accelerations and body roll angles obtained in course of the computer-aided simulation are provided in the graphical representations below (Figs. 2, 3, 4 and 5).

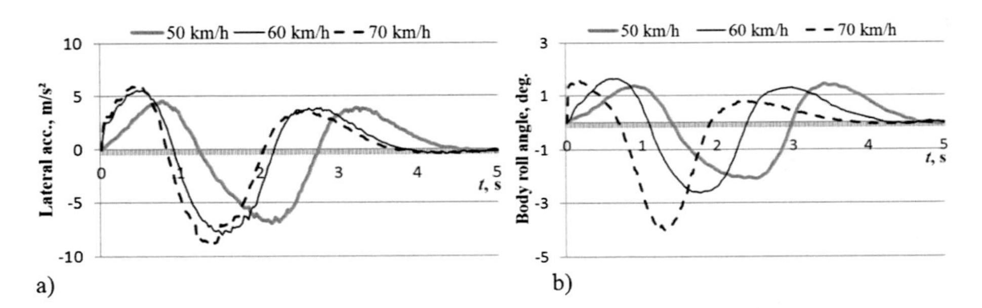

Fig. 2. VW Polo measured characteristics: (a) lateral acceleration; (b) body roll angle.

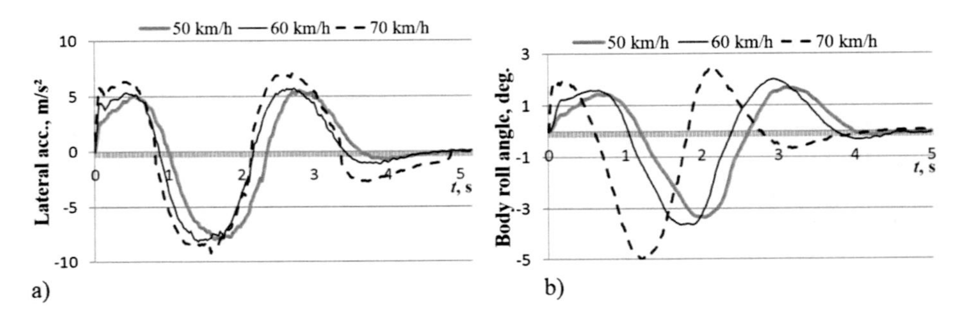

Fig. 3. Seat Ibiza measured characteristics: (a) lateral acceleration; (b) body roll angle.

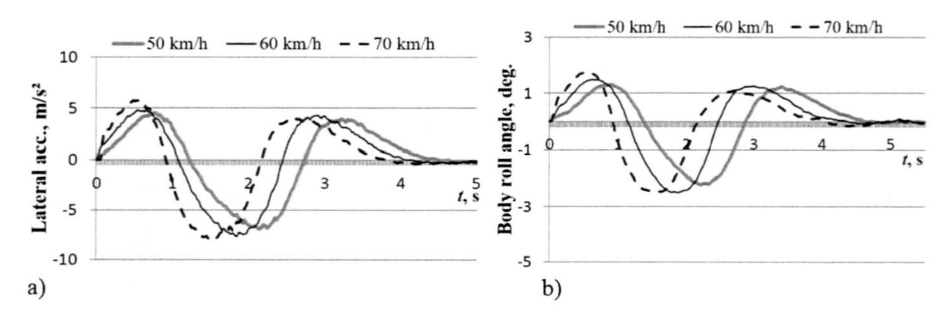

Fig. 4. Ford Fiesta measured characteristics: (a) lateral acceleration; (b) body roll angle.

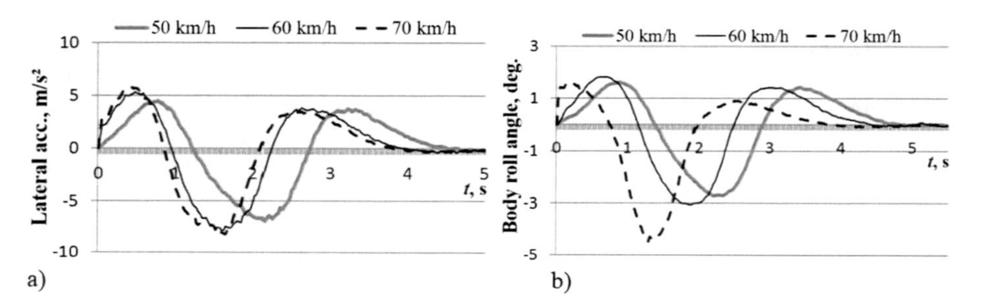

Fig. 5. Kia Rio measured characteristics: (a) lateral acceleration; (b) body roll angle.

It can be seen from the results obtained during the computer-aided simulation that the maximum lateral acceleration was achieved for car Seat Ibiza: it attests the best grip of the wheels with the pavement, although the driver, the passengers and the cargo are affected by the maximum load.

4 The Methodology of the Experimental Research

Both the computer-aided simulation and the experimental research were carried out on dry asphalt pavement using four compact class cars: Volkswagen Polo, Seat Ibiza, Ford Fiesta and Kia Rio; their technical characteristics are provided in the Table 1 below. In a case of a sudden manoeuvre, the said cars face problems of stability.

During the experiments, the cars ought to perform a double lane-change manoeuvre, otherwise referred to the "Elk test". According to standard ISO 3888-2 [8], the distances between the landmarks (in metres) shall be calculated as follows (Fig. 6):

$$A = 1.1k + 0.25; \tag{7}$$

$$B = k + 1.0; \tag{8}$$

where k – the width of the car with the installed equipment in metres.

Table 1. Technical parameters of the cars.

Make, model	VW Polo	Seat Ibiza	Ford Fiesta	Kia Rio
Engine capacity, cm^3	1.2 (petrol)	1.2 (petrol)	1.0 (petrol)	1.1 (petrol)
Power, kW	51	77	92	55
Torque, Nm	112	175	200	170
Mass, kg	967	1015	941	1041
Length, mm	3970	4052	3950	4045
Width, mm	1682	1693	1722	1720
Height, mm	1485	1445	1481	1455
Wheelbase, mm	2470	2469	2489	2570
Track: front/rear, mm	1463/1456	1465/1457	1493/1480	1521/1525
Wheels dimensions	185/60/R15	185/60/R15	185/55/R15	185/65/R15

A double lane-change manoeuvre is the one of open-loop type when the driver's skills are assessed as well in addition. So, upon striving for uniformity of the experiment for all the cars, the test on all the cars was performed by the same driver experienced in car racing.

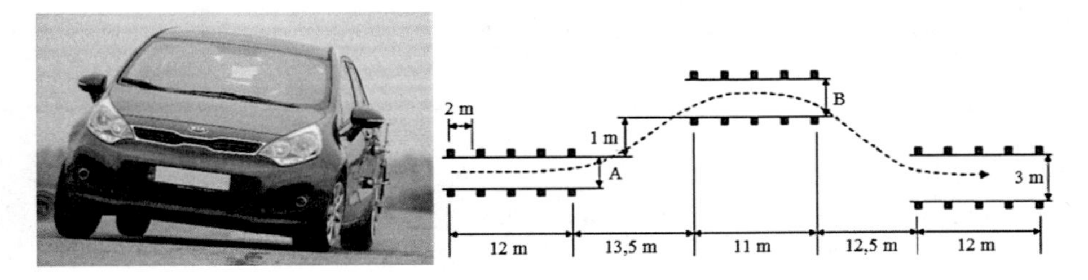

Fig. 6. The column layout of the test (ISO 3888-2) [8].

"The elk test" is carried out by simulation of an obstacle having appeared suddenly on the road. The test starts at the first landmark (Fig. 6) when the vehicle is moving with the constant initial velocity. The said velocity is maintained during the whole test, except of cases when Electronic stability program (ESP) automatically reduces the velocity on sliding of the car. During the test, the driver of the car covers the distance with the constant velocity of 50 km/h and in the section of road equal to 13.5 m, he shifts to the adjacent lane (not reducing the velocity), then moves there 11 m and comes back to the former lane. If the trajectory is successfully covered with the velocity of 50 km/h, the test shall be repeated with the velocities of 60 and 70 km/h upon repeating each test at least twice. During the experiment, the information obtained from the sensors is registered by data collection equipment Corrsys-Datron DAS-3 and later is processed on a mutual time scale by software TurboLab 6.0 [4, 7].

For an objective assessment of a vehicle's stability, the lateral accelerations and the body roll angle affecting the vehicle as well as their changes were measured. Independently on the readings of the vehicle speedometer, the velocities of a car movement in the longitudinal and the lateral direction were fixed individually. For measuring the said parameters, equipment for dynamic testing of vehicles “Corrsys-Datron” was used (Fig. 7). For fixing the dynamic parameters, the frequency of 100 Hz was chosen.

The equipment installed in a car for the tests includes:

- a triaxial accelerometer, the measurement range ±3 g (Fig. 7, position 1);
- a triaxial gyroscope, the measurement range ±150 /s (Fig. 7, position 1);
- velocity measurement sensor Correvit S-350, accuracy < ±0.2% (Fig. 7, position 2);
- two laser-based height change measuring sensors HF-500C, accuracy < ±0.2% (Fig. 7, position 3).

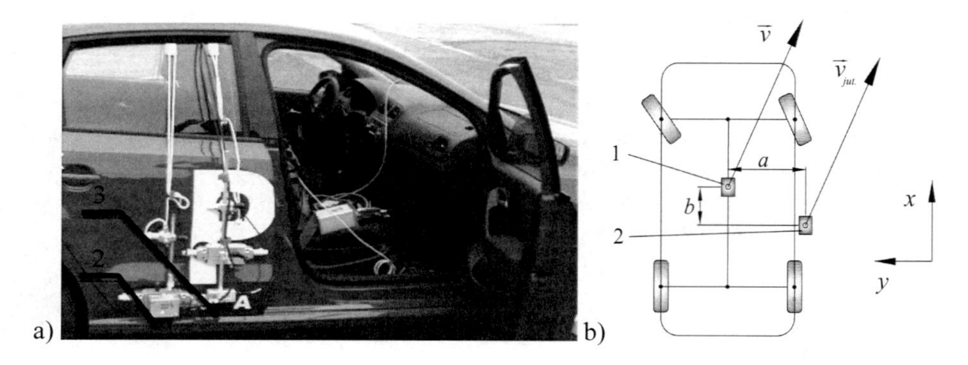

Fig. 7. Vehicle test setup: (a) Experimental research equipment “Corrsys-Datron”; (b) Scheme of measurement: 1 – gyroscope, 2 – velocity sensor.

The velocity sensor was installed in the right part of a car and the gyroscope – close to its centre of gravity. The crew of a car under testing consisted of a driver and a passenger involved in supervision of the measuring equipment.

5 The Results of the Experimental Tests and Their Analysis

During the tests, the lateral acceleration and the body roll angle as well as the intensity of rolling were measured. The achieved maximum values of the said parameters are presented in Table 2 below. These parameters objectively assess the comfort and the freedom of the sprung mass movement; however, stability is directly characterized by the side-slip angle. After a measurement of projections of the longitudinal and lateral velocities by the equipment and adapting a methodology according to them, it is possible to calculate the sliding angle of each car.

According to the scheme for measuring the movement and velocity of a car, the velocity in a vectorial form may be expressed as follows:

$$\vec{v} = (v_{x.}, v_{y.}, 0) = \vec{v}_{jut.} = (v_{jut,x}, v_{jut,y}, 0); \tag{9}$$

where $\vec{v}$ – the generalised vector of velocity; v_x – the projection of the longitudinal velocity; v_y – the projection of the lateral velocity; $v_{jut.}$ – the generalised velocity measured by the sensor.

The velocity sensor evaluates the real velocity vector $\vec{v}$ and the velocity component that appears because of rotation around the vertical axis $\vec{v}_{suk}$ of the vehicle:

$$\vec{v}_{jut} = \vec{v} + \vec{v}_{suk.}; \tag{10}$$

$$\vec{v}_{jut} = (v_{jut,x}, v_{jut,y}, 0) = (v_x - \dot{\psi} \cdot b, 0); \tag{11}$$

After an assessment of the intensity of rotation around the vertical axis and the relative position of sensors, the longitudinal and lateral components of the velocity around the axis of the vehicle will be expressed as follows:

$$v_{x.} = v_{jut,x} + \dot{\psi} \cdot a; \tag{12}$$

$$v_{y.} = v_{jut,y} - \dot{\psi} \cdot b; \tag{13}$$

where a, b – the longitudinal and lateral distance from the gravity center of the vehicle to the sensor (Table 2); $\dot{\psi}$ – the intensity of rotation around the vertical axis of the vehicle (yaw rate).

Then the slip angle may be calculated according to the following formula:

$$\beta = \text{arctg}\frac{v_{y.}}{v_{x.}} = \text{arctg}\frac{v_{jut,y} - \dot{\psi} \cdot b}{v_{jut,x} + \dot{\psi} \cdot a}; \tag{14}$$

The parameters required for the calculation, such as the projection of lateral velocity $v_{jut,y}$, the projection of the longitudinal velocity $v_{jut,x}$, and the velocity of rotation around the vertical axis of the vehicle $\dot{\psi}$ were obtained during the tests from the sensors.

The found values of the side-slip angles $\beta_{\max}$ for vehicles Ford Fiesta and Seat Ibiza are presented in graphic form (Fig. 8).

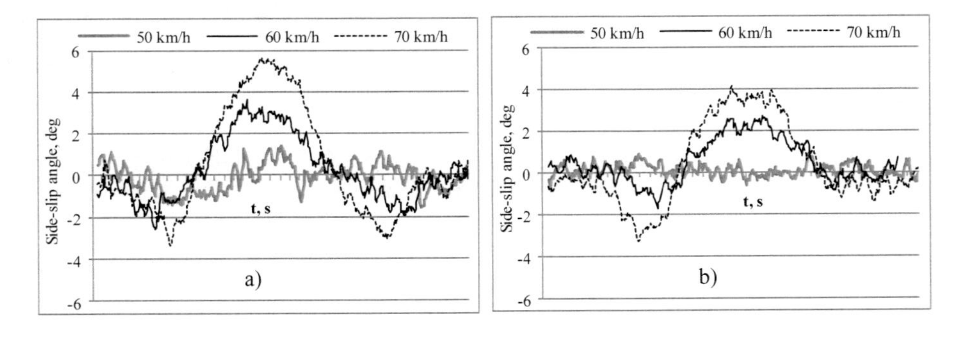

Fig. 8. Evaluated body side-slip angle at different velocities: (a) Ford Fiesta; (b) Seat Ibiza.

The sliding characteristics of the said cars mostly differ. As it can be seen from the curves, the side-slip angle grows with increasing the velocity of the manoeuvre. Even at the velocity of 50 km/h, the maximum angle was found for Ford Fiesta. When the velocity of the manoeuvre was 70 km/h, the side-slip angle exceeded 5.5° and the landmark was clipped.

Table 2. The values of the parameters fixed during the tests.

	a, mm	b, mm	v, km/h	a_y, m/s^2	φ', °/sec	φ, °	β, °
VW Polo	860	1000	50	6.25	11.5	2.9	1.25
			60	7.25	15.0	3.5	1.75
			70	8.25	23.0	4.0	5.5
Kia Rio	880	1020	50	5.5	12.5	2.5	0.5
			60	7.5	22.0	3.5	1.25
			70	8.25	27.0	3.75	4.25
Seat Ibiza	870	1020	50	7.0	13.0	3.25	0.5
			60	7.75	21.0	3.75	2.5
			70	9.5	26.0	4.5	4.0
Ford Fiesta	880	995	50	5.5	11.5	2.5	1.5
			60	7.75	16.0	3.25	3.5
			70	8.25	21.0	3.5	5.75

The reliability of the results of a double lane change manoeuvre is also considerably affected by the driver (his ability to maintain a constant velocity, time of reaction, velocity of steering-wheel rotation). According to the values of the vehicle movement parameters fixed during the tests, it is possible to assess the abilities of the driver to carry out the manoeuvres precisely and in a well-coordinated way. The results of the tests were affected by the type and the properties of the tires as well.

6 Conclusions

- The double lane change manoeuvre had been performed successfully with all the cars used for the tests; however, upon taking into account the values of the slip-angle and the errors that caused clipping the landmarks, it may be stated that the velocity of 70 km/h is the limit to overcome a double lane change manoeuvre for compact class cars.
- The maximum intensity of rotation around the longitudinal axis of the vehicle was achieved for vehicles Seat Ibiza and Kia Rio. This parameter shows the damping characteristics of the suspension – a rapid response to the applied loads. It is important while driving on an uneven road in case of a sudden manoeuvre, because the wheel will be more rapidly pressed to the surface of the road.

- A stability of a car is assessed according to the lateral acceleration, the body side-slip angle and the intensity of rotation around the vertical axis of the car; however, after a correction of the methodology for calculation of the side-slip angle β, it is possible to assess a stability of vehicle movement in details.

References

1. Baffet, G., Charara, A., Lechner, D.: Estimation of vehicle sideslip tire force and wheel cornering stiffness. Control Eng. Pract. **17**, 1255–1264 (2009)
2. Broughton, J.: Car driver casualty rates in Great Britain by type of car. Accid. Anal. Prev. **40**(4), 1543–1552 (2008)
3. Cars-Data.: Cars technical specifications database. http://www.cars-data.com/. Accessed 12 Oct 2018
4. Corrsys–Datron.: Correvit S-350 non-contact 2-axis optical sensor: user manual, vol. 1, 29 p. Corrsys–Datron (2008)
5. Cronje, P.H., Els, P.S.: Improving off-road vehicle handling using an active anti-roll bar. J. Terrramech. **47**, 179–189 (2010)
6. Emirler, M.T., Kahraman, K., et al.: Vehicle yaw rate estimation using a virtual sensor. Int. J. Veh. Technol. **2013**, 13 p (2012)
7. Huber, B., Drews, R.: How to use objective measurement data for Vehicle Dynamics Testing. SAE Int. J. (2009). https://ccc.dewetron.com/dl/52af1137-ea94-419b-80ac-7770d9c49862
8. ISO 3888-2: 2004. Passenger Cars – Test track for a severe lane-change manoeuvre – Part 2: Obstacle avoidance. Geneva (2004)
9. Rajamani, R.: Vehicle Dynamics and Control, 496 p. Springer, New York, Dordrecht, Heidelberg and London (2012)
10. Minghui, L., Yongsheng, Z., Liang, C., et al.: Design of body slip angle observer for vehicle stability control. In: International Conference on Electronic and Mechanical Engineering and Information Technology, pp. 2357–2361. Technology, Shenyang, Liaoning, China, Paris, 7–9 September 2012
11. Pytka, J.A., Trakowski, P., Fijałkowski, S., Budzynski, P., Dabrowski, J., Kupicz, W., Pytka, P.: An instrumented vehicle for off-road dynamics testing. J. Terrramech. **48**, 384–395 (2011)
12. Slaski, G.: Experimental test results of the influence of adaptive damping level on passenger car dynamics during double-lane-change manoeuvre. In: The International Conference of TRANSBALTICA, pp. 200–205. Technika, Vilnius, Lithuania, 5–6 May 2011
13. Uys, P.E., Els, P.S., Thoresson, M.J.: Criteria for handling measurement. J. Terrramech. **43**, 43–67 (2006)
14. Wang, W., Yuan, L., Tao, Sh., Zhang, W., Su, T.: Estimation of vehicle side slip angle in nonlinear condition based on the state feedback observer. In: International Conference on Automation and Logistics. Honk-Kong, Macau, China, 16–20 August 2010
15. Whitehead, R., Travis, W., Bevly, D.M., Flowers, G.: A study of the effect of various vehicle properties on rollover propensity. Auburn University (2004)

The Impact of Legislation to the Traffic Safety

Vigilijus Sadauskas(✉)

Faculty of Transport Engineering, Vilnius Gediminas Technical University,
J. Basanavičiaus g. 28, 03224 Vilnius, Lithuania
vigilijus.sadauskas@gmail.com

Abstract. This paper presents the impact of traffic safety Legislation on Traffic safety. The impact of traffic safety legislation on traffic safety. The traffic safety situation does not improve, as you would like. In recent years, various traffic safety regulations and regulations have been adopted. How many documents, rules, requirements, etc. have been imposed since 2013? How these requirements are influenced the traffic safety situation in Lithuania. What effect was given by a specific requirement? Have you evaluated the effect of these documents, rules and requirements? How often traffic safety needs be changed?

Keywords: Traffic safety · Traffic safety legislation · Amendments · Road Traffic Rules · Road Safety Act

1 Introduction

Single human breast – tragedy, multiple – statistics. As far as I know, for the first time, the word Automobile is mention in 1899 September 14 at the newspaper The New York Time *fatally hurt by automobile* (Fig. 1).

First killed in a traffic accident in Europe in 1896 (Fig. 2).

These Traffic safety strategies are commonly used to describe traffic safety.

The "Triple E" for traffic safety improvements has long been used as a strategy, established perhaps as early as 1930s. The three E s stand for: Engineering, measures enacted in vehicle, road and traffic engineering; Education, training of drivers and traffic education in schools; Enforcement, ensuring and imposing obedience to traffic laws and regulations [1].

R.J. Smeed, an English researcher who showed (in 1949) how road accident fatalities in different countries are related to their population and number of registered motor vehicles. The risk will decrease as the number of motor vehicles per population increases [2].

J.S. Baker, an American researcher pointed out (in 1960) that some chief causes of an accident can be observed in accident investigations but that as a rule there will be no single evident cause of an accidents and systematic analyses are therefore necessary [2].

W. Haddon Jr., an American scientist, set up (in 1970) a framework for categorizing highway safety phenomena, based on three phases Human, Vehicle and Environment [2].

Traffic Exposure Control – the establishment of policies and regulations that reduce transport demand (through organization of the urban fabric, the institutional and

A. Varhelyi et al. (Eds.): VISZERO 2018, LNITI, pp. 188–194, 2020.
https://doi.org/10.1007/978-3-030-22375-5_21

KINGSTON, Jamaica, Sept. 12.—The American schooner Mildred E., from San Blas for New York, stranded here yesterday. She is reported to be a total wreck. The cargo has been salvaged and the crew are going home. Trouble is arising over the wreckers looting the vessel.

FATALLY HURT BY AUTOMOBILE.

Vehicle Carrying the Son of ex-Mayor Edson Ran Over H. H. Bliss, Who Was Alighting from a Trolley Car.

H. H. Bliss, a real estate dealer, with offices at 41 Wall Street, and living at 235 West Seventy-fifth Street, was run over last night at Central Park West and Seventy-fourth Street. He was injured fatally.

Bliss, accompanied by a woman named Lee, was alighting from a south-bound Eighth Avenue trolley car, when he was knocked down and run over by an automobile in charge of Arthur Smith of 151 West Sixty-second Street. He had left the car, and had turned to assist Miss Lee, when the automobile struck him. Bliss was knocked to the pavement, and two wheels of the cab passed over his head and body. His skull and chest were crushed.

Dr. David Orr Edson, son of ex-Mayor Edson, of 38 West Seventy-first Street, was the occupant of the electric cab. As soon as the vehicle was brought to a standstill he sent in a call to Roosevelt Hospital for an ambulance, and until its arrival did all he could to aid the injured man. When he was taken to the hospital Dr. Marny, the house surgeon, said that Bliss was so seriously injured that he could not live.

Smith was arrested and locked up in the West Sixty-eighth Street Station. It is claimed that a large truck occupied the right side of the avenue, making it necessary for Smith to run his vehicle close to the car. Dr. Edson was returning from a sick call in Harlem when the accident happened.

Mr. Bliss boarded at 235 West Seventy-fifth Street. The place where the accident happened is known to the motormen on the trolley line as "Dangerous Stretch," on account of the many accidents which have occurred there during the past Summer.

Fig. 1. Automobile is mention.

Fig. 2. Bridgette Driscoll killed in 1896 (in a circle).

company structure or the daily life of individuals), transfer of a given transport need to safer forms of transport or avoidance of transport using telecommunication instead.

Accident Risk Control – provision of technical measures in Vehicle Road Traffic and Computer engineering directed to eliminate, reduce or detect risks and incidents that can generate accidents (Active safety), and measures that influence road users to behave correctly.

The risk is most conveniently expressed in terms of the number of injured per unit of external factors.

$$R = P_a x C_{a;} \tag{1}$$

where P_a – probability of an accident; C_a – consequences of an accident.

Injury Control – installing protective measures in vehicle design, and softening the road environment such that the consequences of a traffic crash are eliminated or strongly reduced (Passive safety), and efficient rescue service, medical treatment and rehabilitation of persons injured in accidents that cannot be prevented through all measures that have been taken prior to or under the crash situation.

Various tools are used to improve traffic safety: administrative, legislation, rules and law, engineering, construction, educational, scientific, research.

How traffic safety is influenced by legislation policies and rules adopted by politicians?

The behavior of road users is of great importance for traffic safety. In order to make this behavior as uniform and safe as possible, rules established by law on the behavior of road users were developed. In order to ensure compliance with the rules, state authorities exercise control over their execution and punishment for their violation.

Road users, especially car drivers and pedestrians, rate their willingness to participate in traffic much higher than the average for the group to which they belong.

The number of accidents depends not only on how road users behave in the framework of the technical system. To ensure the highest possible traffic safety, legislation on the traffic system should therefore impose requirements not only on road users, but also on government authorities, manufacturers and other persons who form the technical part of the system.

Participants of the movement do not want to behave in accordance with the rules of the movement, if only they have experience in obtaining benefits from violating these rules.

If the rules are reasonable from the point of view of society, they alone do not decide anything. Control and penalties for breaking the rules will always be necessary to ensure respect for the rules established by the state.

The legal regulation of the traffic system aims to ensure the traffic capacity, reduce the number of accidents and distribute responsibility for the control and elimination of risk factors.

Legislation in the field of road traffic forms the legal basis for a number of activities for which the state is responsible.

The rules are addressed to large masses of the population, they should be simple for perception and understanding, so that they are carried out consciously. The number of rules should not be exceptionally large so that road users can be aware of them.

Legal regulation of the behavior of road users and other parts of the road traffic system can reduce the number of accidents only if the following conditions are met: the rules regulate the risk factors, actions or systems that comply with these rules, provide a lower level of risk of accidents than acts or systems that contradict these rules.

The most important factors are:

- how high is the level of knowledge about the rules for those for whom they are intended;
- how likely is it to be detained and punished for breaking the rules;
- how severe and severe are the types of punishment for violations of the rules of the road;

– other advantages or disadvantages are related to the observance or violation of the rules (for example, saving or spending time on a trip, condemning others to your actions, etc.).

The scope of a rule violation is characterized by large variations for individual rules. The level of risk of accidents does not increase as much with all types of rule violations.

2 Safe Traffic System and Its Management Actions

How to calculate the influence of traffic rules or law on the number of accidents? The damage caused by accidents is a major part of the losses incurred by the state [3].

The studies examined the relationship between the number of punishments to which the driver was sentenced for violation of traffic rules and the number of accidents in which this driver was a participant in the same period of time.

All these studies contain the same conclusion: the greater the number of punishable violations of the rules the driver made, the more accidents he was involved in [4].

The basic principle is that a rule violation should be punished in such a way as to comply with this rule. The punishment should be inevitable and effective.

The severity of the punishment must be in compliance with the law. By Ostvik, 1988 the withdrawal of a driving license is perceived as a particularly severe punishment by the vast majority of drivers [2].

When the mandatory withdrawal of a driver's license for 3–6 months was introduced for the first time for drunk driving in the state of Wisconsin, USA, the number of accidents at that time of day decreased by 25% [5].

An investigation into the doubling of the administrative penalty for speeding in Sweden could not detect any change in the frequency of irregularities due to speeding.

Road construction and traffic control are important for the level of risk of accidents. For example, on city streets, the risk of accidents is 10 times higher than the same level on motorways. The risk level at X-shaped intersections is higher than the same indicator at T-shaped intersections, and sharp turns have a higher level of risk than gentle, etc.

The legal regulation of engineering traffic measures has the greatest positive effect, as the number of accidents, injuries and deaths is actually reduced.

The technical requirements for new motor vehicles apply primarily to manufacturers of such vehicles. Over the last 10 years, new requirements have reduced the number of accidents and injuries.

Changes of the Law on Road Traffic Safety from 2013 until 2018. *Adopts the Parliament of the Republic of Lithuania.* What are the benefits of changing the law? Over the past 5 years, more than 25 changes, everytime saying that each change will reduce the number of deaths. Amendments to the law: *"2017-12-21 įstatymas Nr. XIII-974; 2017-10-19 įstatymas Nr. XIII-692, 2018-05-10 įstatymas Nr. XIII-1139; 2016-12-15 įstatymas Nr. XIII-107, 2017-10-19 įstatymas Nr. XIII-695, 2017-12-07 įstatymas Nr. XIII-856, 2017-12-21 įstatymas Nr. XIII-972; 2017-06-22 įstatymas Nr. XIII-491, 2017-11-16 įstatymas Nr. XIII-766, 2017-10-19 įstatymas Nr. XIII-692; 2017-06-01 įstatymas Nr. XIII-403; 2016-12-15 įstatymas Nr. XIII-106; 2015-12-15 įstatymas*

Nr. XII-2186; 2015-12-15 įstatymas Nr. XII-218; 2015-06-23 įstatyms Nr. XII-1849; 2015-05-07 įstatymas Nr. XII-1679; 2015-03-26 įstatymas Nr. XII-1584; 2015-02-06 nutarimas Nr. KT6-N2/2015; 2014-06-26 įstatymas Nr. XII-970; 2014-11-06 įstatymas Nr. XII-1298; 2014-06-26 įstatymas Nr. XII-970; 2014-06-26 įstatymas Nr. XII-969; 2013-06-13 įstatymas Nr. XII-360; 2013-06-13 įstatymas Nr. XII-362".

Changes of the Traffic safety rules from 2013 until 2018. What are the benefits of changing the traffic rules? Over the past 5 years, more than 10 changes, every time saying that each change will reduce the number of deaths. Amendments to the law: *"2018-10-03 nutarimas Nr. 989; 2018-06-20 nutarimas Nr. 598; 2016-11-02 nutarimas Nr. 1077; 2015-11-04 nutarimas Nr. 1135; 2014-10-03 nutarimas Nr. 1086; 2012-11-28 nutarimas Nr. 1436; 2012-06-06 nutarimas Nr. 660".*

Administrative Offense Code in relation to traffic offenses changes from 2013 until 2018. What are the benefits of changing the Administrative Offense Code? Adopts the Parliament of the Republic of Lithuania. Over the past 5 years, more than 5 changes, every time saying that each change will reduce the number of deaths. Amendments to the law: *"2015 m. birželio 25 d. įstatymu Nr. XII-1869; 2016 m. kovo 17 d. įstatymu Nr. XII-2254, 2016 m. lapkričio 8 d. įstatymu Nr. XII-2746, 2017 m. rugsėjo 26 d. įstatymu Nr. XIII-640, 2017 m. lapkričio 16 d. įstatymu Nr. XIII-753, 2017 m. gruodžio 21 d. įstatymu Nr. XIII-973".*

The charts provide the Lithuanian data and changes for the 2013–2018 period at the road safety situation in terms of deaths, injuries and accidents, taking into account the number of Road Traffic Regulations and amendments to the Traffic Safety Act (Figs. 3, 4 and 5).

In reality, the number of deaths in 2013 was 258, in 2014 was 267, in 2015 was 242, in 2018 the number of dead 170. The number of injuries in 2013 was 4,040, in 2014 was 3,785, in 2016 was 3,750, in 2018 the number of injuries 3,738.

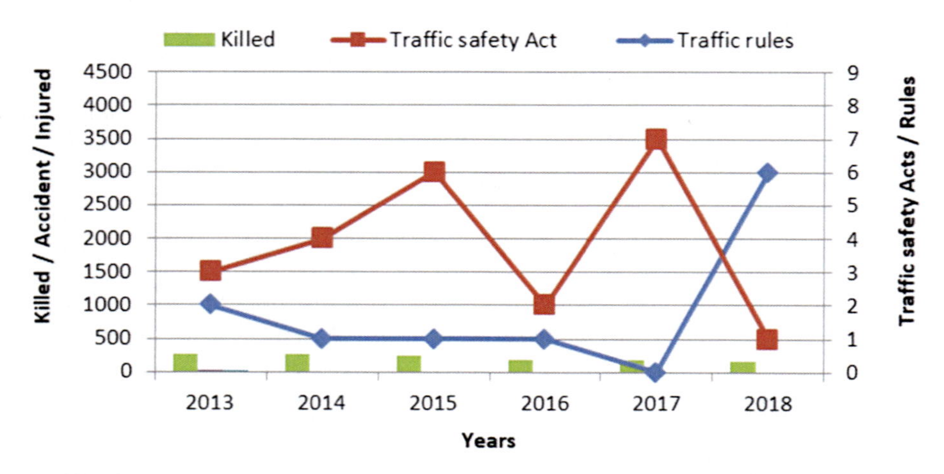

Fig. 3. Killed and amendments to the Road Traffic Rules and the Road Safety Act.

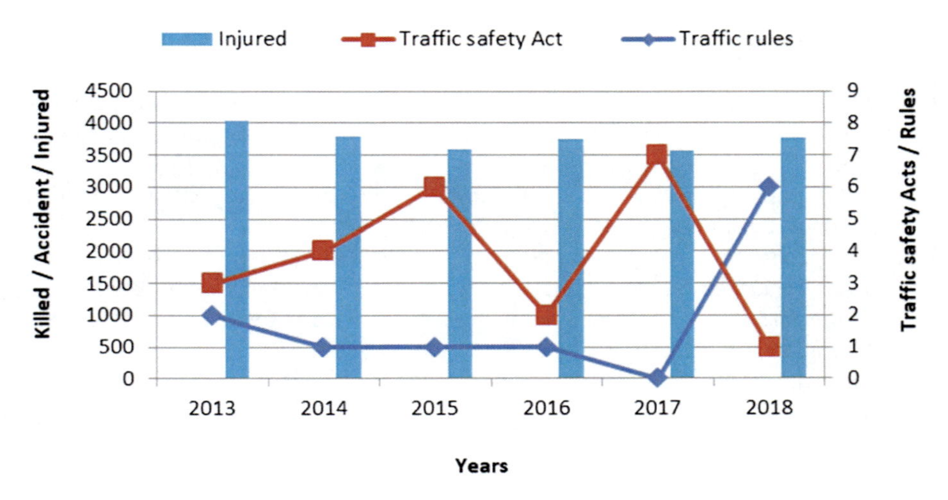

Fig. 4. Injured and amendments to the Road Traffic Rules and the Road Safety Act.

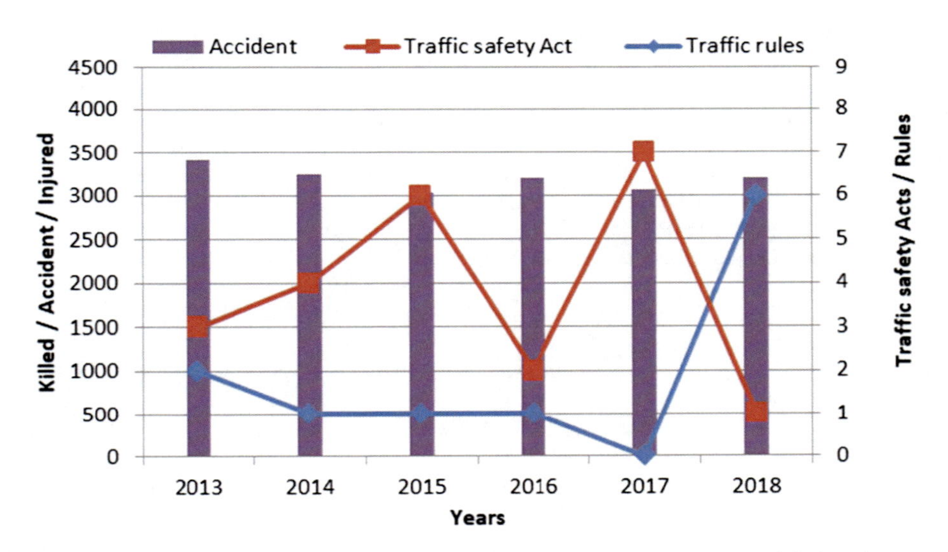

Fig. 5. Accident and amendments to the Road Traffic Rules and the Road Safety Act.

Maybe the legislation is changed only because of the added plus? This research has shown that changes to the Road Traffic Rules and the Law on Road Traffic in Lithuania are not based on real scientific calculations and are not always achieving the desired results.

3 Conclusions

1. If every legislation change would reduce the number of deaths by at least one percent, then the number of deaths would now be the smallest number of deaths in Europe.
2. Over the past 5 years, more than 40 changes, every time saying that each change will reduce the number of deaths.
3. These are changes to only the basic legal acts, there are no ministerial decrees (for example, regarding driver training, examinations, technical inspection, etc.).
4. In reality, the number of deaths in 2013 was 258, in 2014 was 267, in 2015 was 242, in 2016 was 192, in 2017 was 192.
5. There is no analysis what effect has been given or not been given by the specific legislative amendment.

References

1. Sadauskas, V.: Drivers age and traffic accidents in Lithuania//transport means. In: Proceedings of the 15th International Conference, pp. 153–155. Kaunas University of Technology, Technologija, Lithuania, Kaunas, October 20–21, 2011. ISSN 1822-296X
2. Elvik, R., Vaa, T.: The Handbook of Road Safety Measures, p. 1078. Elsevier, Amsterdam (2004)
3. Pukalskas, S., Pečeliūnas, R., Sadauskas, V., Kilikevičienė, K., Bogdevičius, M.: The methodology for calculation of road accident costs. Transport **30**(1), 33–42 (2015). ISSN 1648–4142. eISSN 1648-3480
4. Matthews, M.L., Moran, A.R.: Age differences in male drivers' perception of accident risk: the role of perceived driving ability. Accid. Anal. Prev. **18**, 299–313 (1986)
5. Preusser, D.F., Blomberg, R.D., Ulmer, R.G.: Evaluation of the 1982 Wisconsin drinking and driving law. J. Saf. Res. **19**, 29–40 (1988)
6. Register of Legislation.: https://www.e-tar.lt/portal/lt/legalAct/2ecfa200ec8e11e78a1adea6fe72f3c5. Accessed 10 Oct 2018
7. Register of Legislation.: https://www.e-tar.lt/portal/lt/legalAct/178d4200cacd11e8bf37fd1541d65f38. Accessed 10 Oct 2018
8. Register of Legislation.: https://www.e-tar.lt/portal/lt/legalAct/c28917a0790211e8ae2bfd1913d66d57. Accessed 10 Oct 2018
9. Register of Legislation.: https://www.e-tar.lt/portal/lt/legalAct/e84b03b0a19811e69ad4c8713b612d0f. Accessed 10 Oct 2018
10. Register of Legislation.: https://www.e-tar.lt/portal/lt/legalAct/a0afbf6083c011e5b7eba10a9b5a9c5f. Accessed 10 Oct 2018
11. Register of Legislation.: https://www.e-tar.lt/portal/lt/legalAct/45a8c2f0506e11e4a698d921e3e46801. Accessed 10 Oct 2018
12. Register of Legislation.: https://www.e-tar.lt/portal/lt/legalAct/TAR.B61B5D4CF90B. Accessed 10 Oct 2018
13. Register of Legislation.: https://www.e-tar.lt/portal/lt/legalAct/TAR.400F0B4A900C. Accessed 10 Oct 2018
14. Register of Legislation.: https://www.e-tar.lt/portal/legalAct.html?documentId=dcab42a026dc11e5bf92d6af3f6a2e8b. Accessed 10 Oct 2018

Evaluation of Shopping Mall Implementation Impact on Safety Aspect of the Transport Network Based on Simulation: Case-Study of Riga

Mihails Savrasovs(✉)

Transport and Telecommunication Institute,
Lomonosova str. 1, 1019 Riga, Latvia
savrasovs.m@tsi.lv

Abstract. Implementation of new urban attraction point usually leads to the changes in the surrounding transport infrastructure, which could impact on the performance of the transport network, impact on environment quality etc. Mentioned aspects could be evaluated using microscopic traffic flow simulation tools, which could provide a set of measures to perform the evaluation of the impact. Usually, the impact on safety aspects in a transport network is not evaluated, as it is not requested by the local municipality rules (at least in Latvia). But at the same time, it is an important issue, which has a direct impact on the transport network reliability and safety of drivers and pedestrians. The current paper aim is to demonstrate the case-study of microscopic traffic flow simulation model application for evaluating the surrogate safety after introducing the new urban attraction point (shopping mall) and changes in the surrounding transport infrastructure. The paper defines the methodology for evaluating the surrogate safety, based on microscopic traffic flow simulation model and presents the evaluation results of surrogate safety, completed for one of the newly constructed shopping mall in Riga city (Latvia). Based on the obtained data the conclusions about the impact of new urban attraction point are presented.

Keywords: Traffic flow simulation · Shopping mall · Safety

1 Introduction

Implementation of new urban attraction point usually leads to the changes in the surrounding transport infrastructure, which could impact on the performance of transport network, impact on environmental quality, impact on traffic flows circulation in the specific area, safety etc. Evaluation of all above-mentioned aspects usually is done by different consulting companies, which apply different methodologies and usually not interrelated. As the core tool for evaluating different aspects related to urban attraction point development, it is proposed to use traffic flow microsimulation approach, which provides the ability to do detailed traffic flow simulation in the study areas. It was demonstrated in previous publications, that the model of transport network

A. Varhelyi et al. (Eds.): VISZERO 2018, LNITI, pp. 195–205, 2020.
https://doi.org/10.1007/978-3-030-22375-5_22

could be used not only to evaluate the performance of the existing and planned transport infrastructure but also to measure different aspects, like as example evaluation of the accessibility of the new attraction point [8], evaluation of the environment quality [7]. So, it seems reasonable also to apply the developed models to evaluate the safety of the planned transport infrastructure. The idea of the microscopic transport models use for this purpose is not new and there are publications, which proposes different approaches and methods, like as example in Wang et al. [11] there is discussed a new model for evaluating safety, based on microsimulation, Tan et al. [9] there is a proposed model for evaluating safety in signalized intersection, and finally initial idea was discussed in Archer and Kosonen [1]. Most of the publication presents new models for safety evaluation, but as a general disadvantage of all of them are lack of practical application and presentation of real case studies.

Usually, the objective safety level of transport system could be measured by the number of police-reported accidents and the severity of their outcomes in terms of personal injury and fatality. But this approach gives only empirical data, which is not possible to apply the data for evaluating the safety level for the planned transport infrastructure.

The goal of this publication is to present the case-study for evaluation of shopping mall implementation impact on the safety aspect of the transport network and to discuss the obtained results.

2 Case-Study Description

Current publication deals with the case-study which is targeted on evaluating the safety level before and after implementation of the shopping mall in a specific district or Riga city, the location of the shopping mall is represented in Fig. 1 as the red circle.

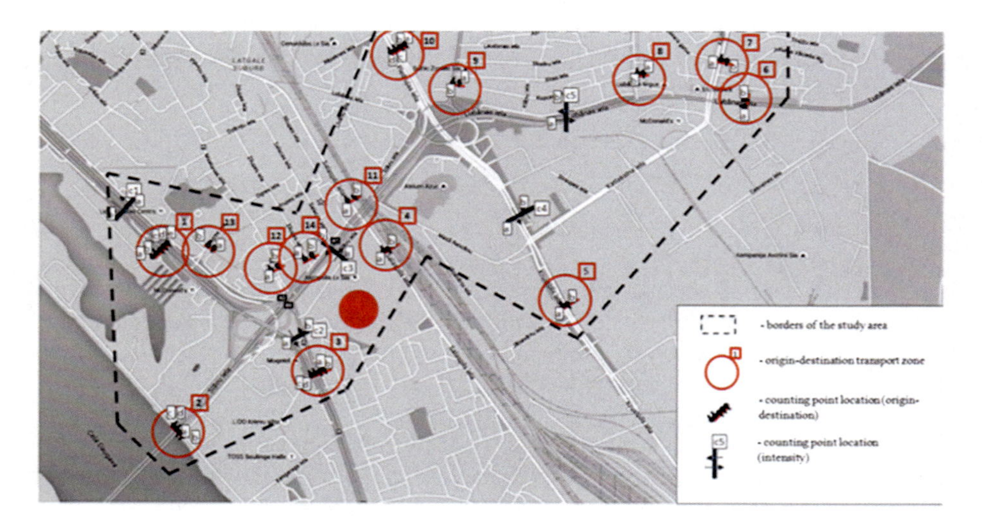

Fig. 1. Study object (background from Google Maps).

To evaluate the impact of safety level the fragment of the transport network around the shopping mall was selected as the study object. The study object could be treated as a complex urban transport node as it has a significant number of signalised and not signalised crossroads, three-level flyovers. Moreover, the selected fragment of the transport networks plays a significant role in connecting the city central area and residential districts. To note, that area around the shopping mall relates to major city roads, highly used during peak hours.

2.1 Current State of the Network

As was mentioned earlier the area of the shopping mall is linked with major urban roads of the city. Figures 2 and 3 demonstrates the most congested parts of the network during morning and evening peak hours. Based on data it could be concluded that evening peak hour is providing more impact on transport network compare to morning peak hour. So, based on the obtained data about congestion level it was made a decision to use only data for evening peak hour in the research. In addition, it is important to note, that it is expected, that a new shopping mall will attract more traffic flow by the evening time.

Fig. 2. Transport network state during morning peak hour (data from Google Maps).

Fig. 3. Transport network state during evening peak hour (data from Google Maps).

2.2 Input Data for Simulation

To reach the goal of the research it was decided to use microscopic traffic flow simulation approach as a tool for obtaining data about traffic movement in the area, before and after the new shopping mall implementation. For simulation model development the regular input data were used as follows: actual demand data; validation data; forecasted demand data; actual and planned geometrical and controlling signs data. More details regarding input data for simulation could be found in Savrasovs et al. [8].

3 Base Model Development

For simulation model development of the fragment of the transport network PTV VISSIM simulation software was applied. In process of model development, the standard model development methodology was used, which is consist of several major steps as follows: modelling of the transport network and its properties, definition of the traffic flows, definition of traffic flow movement rules and regulators, configuration of the output data, model animation development.

3.1 Coding of Model Supply Data

Based on area topography data transport network was coded using standard software tools (links & connectors). The Fig. 4 represents the implemented network. Additionally, important transport infrastructure elements like public transport stops, signal controllers, signal heads etc. Based on open data the public transport lines and schedule of public transport was coded. To define the priorities for traffic flows the conflict area objects were used in the implemented model. The coded network was tested to avoid technical mistakes in the implementation of this part of the simulation model.

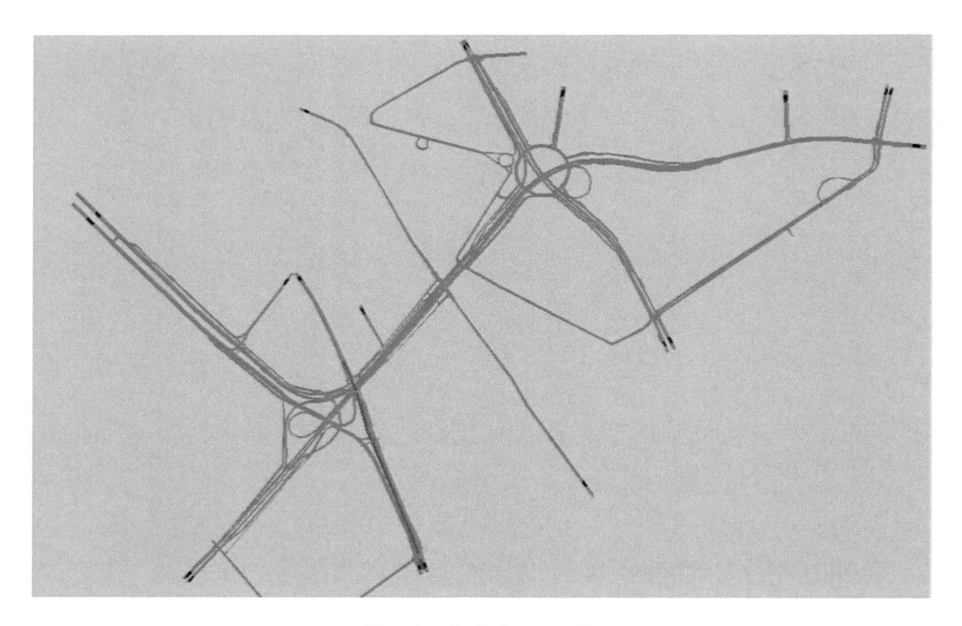

Fig. 4. Coded network.

3.2 Definition of Model Demand Data

The demand part of the models represents the volume and structure of the traffic flow. In this case-study, the demand was coded as a set of OD (origin-destination) matrices one per each vehicle's type. The coding of demand data was done using the regular approach of the VISSIM software, meaning that parking lot objects were used to define the origin and destination of each transport zone. So, in total fourteen transport zones were coded in the model. So, the dimension of OD matrices was 14 × 14. The data in OD matrices describe the volume of transport from origin to destination for 1 h 15 min, as 15 min was used as a warming-up period. Value of the warming-up period was evaluated as the doubled time for vehicles to travel through the network.

3.3 Initial Model Validation

Before model use for impact evaluation of a new shopping mall into safety of the transport network, it is necessary to complete model validation and calibration. Both were done for the implemented model. Results from the final step of the validation are presented in Table 1.

Table 1. Validation results summary.

Observation point	Direction	Simulated values	Counted values	Difference (%)	GEH
C1	A	2658	2804	5.2	2.71
	B	4649	5016	7.3	5.27
C2	A	3003	3222	6.8	3.90
	B	1846	1967	6.2	2.77
C3	A	4626	4909	5.8	4.09
	B	4960	5267	5.8	4.29
C4	A	1297	1379	5.9	2.24
	B	1136	1223	7.1	2.53
C5	A	1562	1658	5.8	2.39
	B	2157	2302	6.3	3.07

Table 1 gives a general view on obtained validation results, which were obtained after the iterative procedure of model calibration. Simulated values refer to the data obtained from the model using standard output data collection tools and represent the volume of traffic passing the specific point of the network during simulation period, while counted values refer to the data collected for validation purpose in specific places of the transport network. The location of counting points is represented in Fig. 1. as bold black lines. To make a decision about developed model validity GEH (Geoffrey E. Havers) index was used. The GEH index value was evaluated for each traffic counting point and calculated as follows:

$$GEH = \sqrt{\frac{2(M - C)^2}{M + C}} \tag{1}$$

where M – simulated traffic volume; C – actual traffic counts.

According to the GEH index use methodology, 85% of the volumes in a model should have GEH less than 5. From Table 1 could be seen, that only one value (for one counting point and one direction) is higher. Additionally, to GEH index, the NAÏVE approach was used to make the decision about model validity. NAÏVE approach according to [6] is based on linear regression model construction using two variables simulated volume and actual volume of traffic. As quality index regular regression model quality characteristics were used R^2, RMSE. To conclude, the developed model could be treated as valid and could be used for further research. More details about validation of the model and more deep analysis of the model could be found in Savrasovs et al. [8].

4 Future Model Development

To implement future transport network and traffic development the model of the current state was used as a base. The base model was calibrated in a previous stage to fit the current state of the traffic. The model was updated from point of view of the transport network and traffic data.

4.1 Transport Network Development

The development of the shopping mall is linked with the significant update of the surrounding transport infrastructure. Figure 5 shows the updated transport network, which was coded in VISSIM. Based on shopping mall development master plan data, 6 additional traffic light signal controllers were added to the model. The parking places were described in the model using Parking objects in the abstract mode and with linking them to the additional transport zone (Transport zone No. 15 for a shopping mall). To simulate the parking process in the area of the new shopping mall the software object – abstract parking place was used to describe parking lots. Parking lots were presented in abstract mode, meaning, that not individual parking lots were coded in the model, but only the capacity of each parking place was defined based on shopping mall master plan.

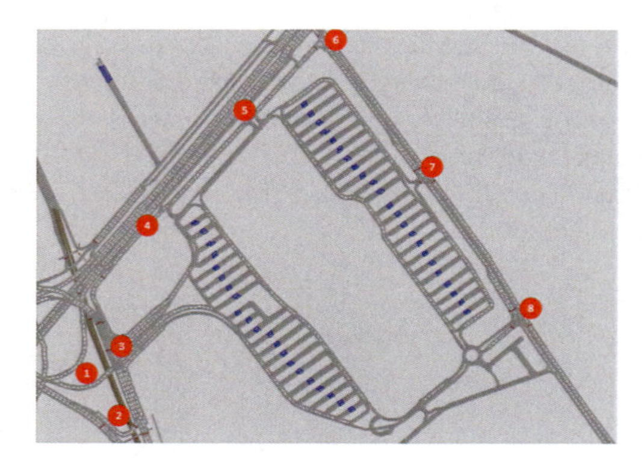

Fig. 5. Key changes in the model compare to the base model.

Additionally, the speed restrictions were organised in the territory of the Shopping mall. The speed limit was defined as 20 km/h (as an average value of speed with deviation up to 5%).

Figure 5 represents the general view on the updated base model, marking key changes in the transport network, which are described in detail in Savrasovs et al. [7].

4.2 Future Demand

The data about demand were organized as OD matrices: Matrix of Friday Peak's Current State Trips (reduced with Pass-By Trips), Urban Primary Trips, Matrix of Friday Peak's Regional Primary Trips. The Pass-by Trips were organised in form of trip chain file in specific VISSIM format, defining the origin of travel, the destination of the travel and time spent in the shopping mall, which was evaluated as random value distributed by normal distribution law with average 1 h and deviation of 15 min.

5 Evaluation of Safety

As a core tool for evaluating the safety aspect of the initial and future transport network, it is proposed to use Surrogate Safety Assessment Model (SSAM) provided by Federal Highway Administration of USA [3]. The SSAM is widely used and several examples could be found in Ghanim and Shaaban [4], Vasconcelos et al. [10], Kim et al. [5] etc.

According [3] to evaluate transport network using SSAM, the transport network should be simulated. Simulation provides the trajectory data pet vehicle in the model. Next, SSAM is used as a post-processor to complete of analysis of trajectory data, usually obtained not from one simulation run, but from several.

SSAM model does vehicle-to-vehicle interactions analysis and identifies conflict events, based on internal conflict model. For each such event, SSAM also calculates several surrogate safety measures, including the following:

- Minimum time-to-collision (TTC).
- Minimum post-encroachment (PET).
- Initial deceleration rate (DR).
- Maximum deceleration rate (MaxD).
- Maximum speed (MaxS).
- Maximum speed differential (MaxDeltaV).
- Classification as lane-change, rear-end, or path-crossing event type.

In SSAM are defined following 3 types of possible conflicts: rear-end, lane-change and crossing (see Fig. 6).

The general workflow of safety evaluation could be presented as the procedure demonstrated in Fig. 7. The procedure includes steps necessary to obtain the safety measures of SSAM. As could be seen from Fig. 7 before application of SSAM the model should be developed, calibrated and validated. The results of model development, calibration and validation are presented in previous sections of the paper.

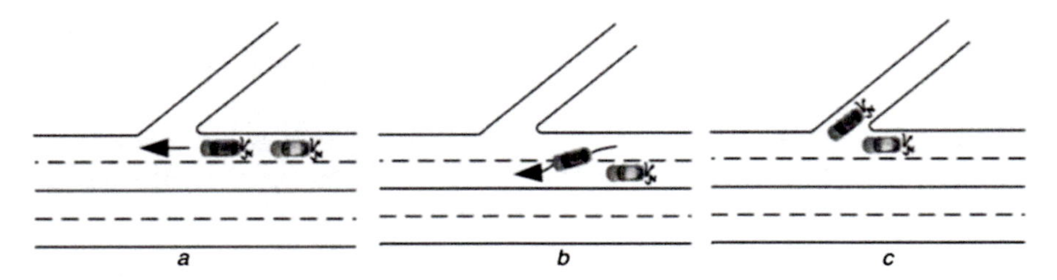

Fig. 6. SSAM conflict types (a: rear-end conflict; b: lane-change conflict; c: crossing) [2].

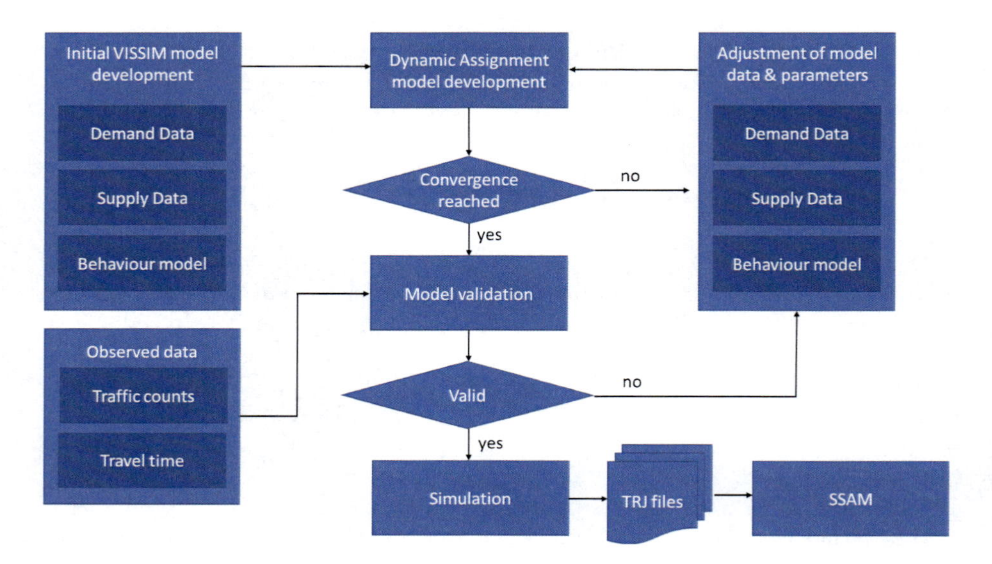

Fig. 7. General workflow for safety evaluation.

The significant affords of model development and validation were related with the use of dynamic assignment in the model. This added the additional step of checking the convergence criteria (Travel on paths time difference <15%).

After model validation, the TRJ files were generated and processed for each scenario separately. The results of the SSAM output are presented below. Figure 8 demonstrates statistical data about the distribution of identified conflicts by type for 1st scenario (without shopping mall) and 2nd scenario (with shopping mall). As could be seen from Fig. 8 the majority part of identified conflicts is related to rear-end conflicts, while the second largest group is lane-change conflicts. While it is observed the growth of the number of crossing conflicts, after implementation of the shopping mall, it is related with new ramps introduction in the network.

Fig. 8. Distribution of conflicts by type for 1st and 2nd scenario.

The number of identified conflicts for the 1st scenario is 77204, for the 2nd scenario is 83809. A large amount of the identified conflicts is related with the complexity of simulated transport node and with new parking area near the shopping mall.

The Table 2 below represents SSAM measures produced after trajectory data files processing.

Table 2. Surrogate safety measures.

Measure	1st scenario			2nd scenario		
	Min	Max	Mean	Min	Max	Mean
TTC	0	1.5	1.05	0	1.5	1
PET	0	4	2.78	0	4	2.62
MaxS	0	22.34	4.72	0	22.53	4.88
DeltaS	0	41.86	3.26	0	42.48	3.45
DR	−8.35	3.5	−3.04	−8.35	3.5	−2.85
MaxD	8.45	3.5	−5.12	−8.48	3.5	−4.78
MaxDeltaV	0	29.28	1.78	0	27.31	1.89

The application of t-test demonstrated, that most of the surrogate safety measures do not have a significant difference (tested using a t-test). While the comparison of spatial distribution (see Fig. 9) shows, that parking near shopping mall and ramps are becoming another source of conflicts. But in the same time, it could be seen, that the improvement of the transport infrastructure and attraction of pass-by vehicles to shopping mall gave also a positive result, meaning, that some volume of vehicles attracted by the shopping mall, and they are not participating movement during peak hours in the selected area.

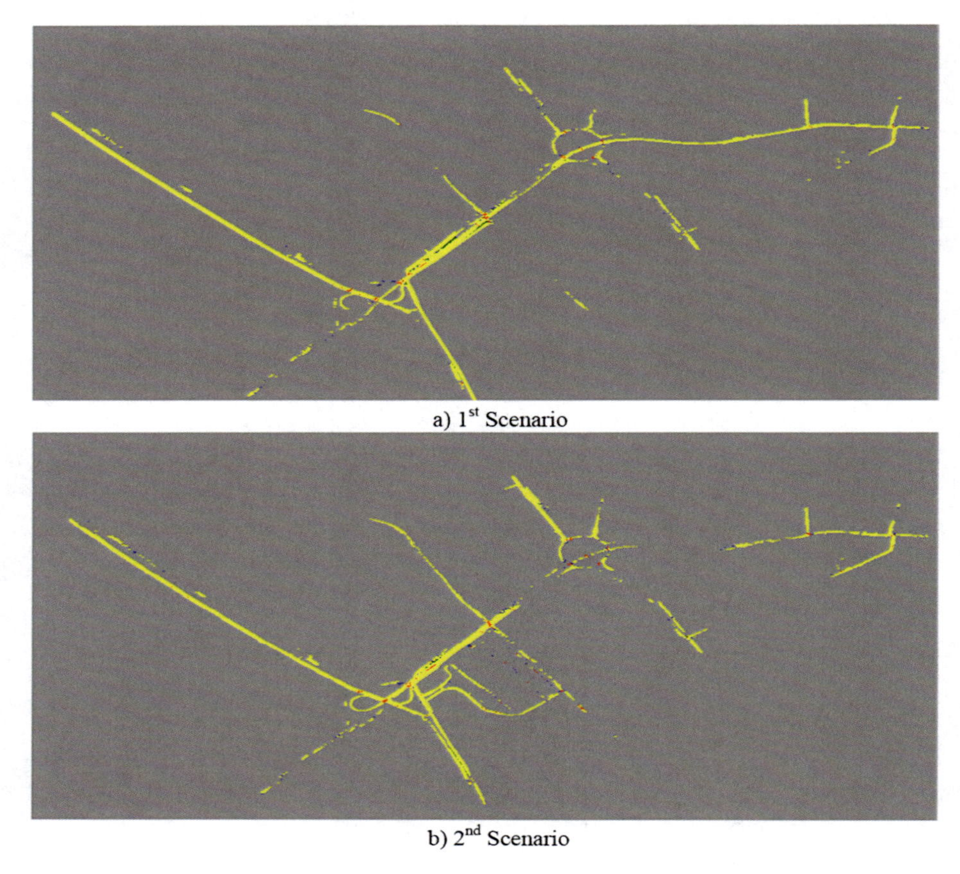

a) 1st Scenario

b) 2nd Scenario

Fig. 9. Spatial distribution of conflicts by type (yellow: rear-end; blue: lane change; red: crossing).

6 Conclusions

Current publication aim is to provide details about case-study of applying microscopic traffic simulation for evaluating safety measures. Usually, local municipality requires to provide traffic impact analysis results in case of implementation of new attractions point in the urban area. But usually simulation results regarding traffic measures are provided, at the same time environmental impacts and safety impacts are provided separately by different consulting companies who apply different methodologies and approaches. This publication demonstrates, that the traffic simulation model could be the core of the analysis and could be used to evaluate safety measures. As the core of safety measures evaluation SSAM is used as a tool for processing the model trajectory files. SSAM classifies all conflicts as rear-end conflicts, lane-change conflicts and crossing conflicts. Mentioned conflicts are major conflicts which lead to accidents in the urban area. The analysis of data obtained for both scenarios demonstrated the growth of conflicts number and a percentual increase of the number of crossing conflicts in the 2nd scenario. This could be explained by the additional number of rams, which will be constructed to access the shopping mall area. In the same time more, specific analysis of the measures did not demonstrate a significant difference (was checked using a t-test).

References

1. Archer, J., Kosonen, I.: The potential of micro-simulation modeling in relation to traffic safety assessment. In: Proceedings of the ESS Conference 2000, Conference Presentation, Germany (2000)
2. Fan, R., Wang, W., Liu, P., Yu, H.: Using the VISSIM simulation model and surrogate safety assessment model for estimating field measured traffic conflicts at freeway merge areas. IET Intel. Transp. Syst. **7**(1), 68–77 (2013)
3. Fhwa.dot.gov: Surrogate Safety Assessment Model (SSAM) - FHWA-HRT-08-049. https://www.fhwa.dot.gov/publications/research/safety/08049/. Last accessed 1 Jan 2019
4. Ghanim, M., Shaaban, K.: A Case study for surrogate safety assessment model in predicting real-life conflicts. Arab. J. Sci. Eng. **44**, 4225–4231 (2018)
5. Kim, K., Saito, M., Schultz, G., Eggett, D.: Evaluating safety impacts of access management alternatives with the surrogate safety assessment model. J. Transp. Res. Board **2672**, 120–128 (2018)
6. Kleijen, J.P.C.: Validation of models: statistical techniques and data availability. In: Winter Simulation Conference 7 (1999)
7. Savrasovs, M., Karakikes, I., Pticina, I.: Shopping mall environmental impact evaluation based on microscopic traffic flow simulation. information modelling and knowledge bases XXX. In: Proceedings of the 28th International Conference on Information Modelling and Knowledge Bases, {EJC} 2018, Riga, Latvia, 4–8 June 2018, pp. 125–136 (2018)
8. Savrasovs, M., Pticina, I., Zemlynikin, V.: Shopping malls accessibility evaluation based on microscopic traffic flow simulation. In: Nathanail, E.G., Karakikes, I.D. (eds.) Advances in Intelligent Systems and Computing Data Analytics: Paving the Way to Sustainable Urban Mobility, pp. 856–863. Springer Nature Switzerland, Greece (2019)
9. Tan, D., Alhajyaseen, W., Asano, M., Nakamura, H.: Development of microscopic traffic simulation model for safety assessment at signalized intersections. J. Transp. Res. Board **2316**(1), 122–131 (2012)
10. Vasconcelos, L., Neto, L., Seco, Á., Silva, A.: Validation of the surrogate safety assessment model for assessment of intersection safety. J. Transp. Res. Board **2432**(1), 1–9 (2014)
11. Wang, C., Xu, C., Xia, J., Qian, Z., Lu, L.: A combined use of microscopic traffic simulation and extreme value methods for traffic safety evaluation. Transp. Res. Part C Emerg. Technol. **90**, 281–291 (2018)

Drowsiness in Drivers of Different Age Categories While Performing Car Following Task

Alina Mashko([✉]) and Adam Orlický

Czech Technical University in Prague,
Konviktská 20, 110 00 Prague, Czech Republic
alina.mashko@gmail.com

Abstract. Decrease of vigilance caused by sleep deprivation affects driving behaviour and results in delay in reaction and loss of vehicle control. This behaviour leads to increase of road accidents. The effect is typical at long commutes and monotonous roads with tedious landscape and is especially noticeable in so called vulnerable diver population groups, namely novice and senior drivers, shift and overtime workers. Prediction of such a behaviour shall contribute to vehicle safety systems through improvement of driver-car interface for prevention of sleepiness at wheel and decrease number of accidents caused this factor, where timely detection of a driver loss of vehicle control is crucial. Current study represents an experimental research on driving simulator with sleep deprived subjects from two age groups (senior, experienced drivers and young, novice drivers) and compares the vehicle control and lateral position ability based on analysis of vehicle lane position, deviation of speed, lane departure time and lane departure area in subject's two state (fresh, rested state and induced sleep deprived state), while performing a car following task with speed-change cycles.

Keywords: Driver drowsiness · Sleep at wheel · Car following · Driver sleepiness

1 Introduction

Drowsy driving is caused by numerous external socio-physiological factors, such as insufficient night sleep, excessive daytime sleepiness, start and duration of work shift, family status, physical and mental workload, individual circadian rhythms, body mass index, use of illicit drugs and alcohol to name a few [1, 2]. Research proves reasonable awareness of drivers of signs of sleepiness [3], however mitigation or prevention of the causes is hardly accessible especially inside the contemporary vehicle and technology currently available at the market which is currently focused at dealing with the consequences itself – the sleep at wheel. A broad review of such systems is available in [4]. Early generation of in-vehicle sleep detection is in a way established based on individual countermeasures, such as roadside sleepiness for consuming caffeine, 'stretching' one's legs, having a rest or nap break including the less efficient ones such as listening to music, opening window etc. [3]. Nonetheless, the same research indicates

A. Varhelyi et al. (Eds.): VISZERO 2018, LNITI, pp. 206–214, 2020.
https://doi.org/10.1007/978-3-030-22375-5_23

that personal awareness in almost 70% cases is not sufficient for prevention individuals from continuing driving in a drowsy or sleepy state.

A significant percentage (20% worldwide) of fatal accidents is connected to driver fatigue [5]. According to accident type, crashes classified as related to fatigue include single vehicle crashes, hitting a fixed object and rollover [6, 7].

Young drivers (under 30 years old) are highly subjected to driving while fatigued and, consequently, are more frequently involved in fatigue-related accidents, which is explained by their lifestyle and tendency to drive later at night, where the most influencing factor in relation to accidents is inexperience [8, 9].

The lifestyle and energy consuming schedules, such as overtime work, household obligations as well, which are all are an adding factor to being exposed to fatigue, which contributes to drowsy driving behaviour for middle aged drivers. Moreover, due to health issues and sleep problems, senior subjects are highly subjected to fatigue, though were found to be less eager to drive in the state of drowsiness [10].

2 Related Work

Driver fatigue, drowsiness and sleepiness research applies the approach of different fields, including psychology, engineering, neural science. When analysing the driver state, it is common to apply complex assessment methodology that may include subjective assessment or self-assessment, biological measures and technical measures. The subjective methods of assessment include implementation of grading scales, questionnaires, self-assessment, expert appraisal and others, which are described in more detail in previously published research [11]. Biological measures for drowsiness include those related to subject physical state and include assessment of brain activity with application of EEG, eye behaviour and movements with application of EOG and eye-tracking devices, measurement of heartbeat (ECG), respiration rate, while most of the measurement devices are not contactless. As a non-invasive technology, camera-based systems tend to be the most promising technology for application in technology, provided the detection is at an automatic level. Objectively, it is possible to assess the driving behaviour based on the vehicle measures, which includes the speed, use of break and acceleration pedals, lane position variability, steering and reaction times. It is typical to conduct driver drowsiness related research in driving simulator laboratory, as it includes observation of impaired state caused by induced sleep deprivation prior to the experiment which would affect road safety. Some related research has been implemented in real traffic implied a real driving followed normal night rest of participants [12], or under supervision [13]. Significant variability in speed and lateral position was observed between daytime and night time driving, where in the last case lower speeds and driving to the midline were recorded [12]. Besides others, lane departure seems to be one of the typical features of falling asleep while driving (e.g., the research in [14]), therefore the lane position and its deviation is one of the important measures. Lane position standard deviation (SD) was found as corresponding to subjective sleepiness rating according to Karolinska Sleepiness Scale [15], while the increase in the deviation tends to increase during the first hour of ride [16]. The lane deviation (and consequent lane exit) may stand both for vehicle control loss as well as to steering correction movements, which corresponds to increase in steering angle as the result of lower accuracy of steering [17].

A longitudinal analysis for detection of vigilance loss due to drowsiness and fatigue shows deviation: increase in time headway with respect to subjective KSS and eyelid closure PRECLOS, while mean value of headway has shown negative correlation to PRECLOS [18].

3 Research Method

The experimental research is focused on analysis of driver of two age groups experimental.

3.1 Testing Cohort

The research was performed on 12 male subjects of two age groups: young/novice drivers: $n = 8$, $s = 3.6$, $s_2 = 12.98$, $SD = 3.37$, $SD_2 = 11.36$, mean = 23.87; elder/experienced drivers $n = 4$, $s = 13.96$, $s_2 = 194.6$, $SD = 12.09$, $SD_2 = 146.19$, mean = 49.75. All subjects were measured in two states: rested (baseline) state with at 6.5 + 2.5 h nigh sleep and after 24 + 1.5 $\sim$ 6 h of night sleep deprivation, while their sleep during the night before measurement was 0 + 1.5 h. The measurements were run in the morning, or in the afternoon. All subjects were instructed not to take caffeine containing drinks and energy stimulants for 24 h and alcohol 72 h prior to measurement.

3.2 Testing Setting and Scenario

The scenario used for measurement was represented by a highway with tedious landscape, minimum curvature driving track and low traffic in both directions for provoking the sleepiness incidents and included the leading vehicle, which per the instruction subjects were following during rides (see Fig. 1). The leading vehicle drives according to the predefined changing speed cycles in a range between 75–125 km/h and are schematically represented in Fig. 2.

The experiment was run on a steady-based driving simulator with half-cockpit of Škoda Octavia cut in back of driver seat.

3.3 Data Evaluation

The data collected from vehicle simulator provided information of driving behaviour based on lane position that included measures within one drive of each subject: average deviation of vehicle from mid line (Av.mld), lane departure count (NLE), average deviation from leading vehicle speed (ASD), lane departure time (TLD) and the lane departure area (LDA). The measures were compared between subjects two states and between two age groups.

Fig. 1. A leading vehicle from driving scenario.

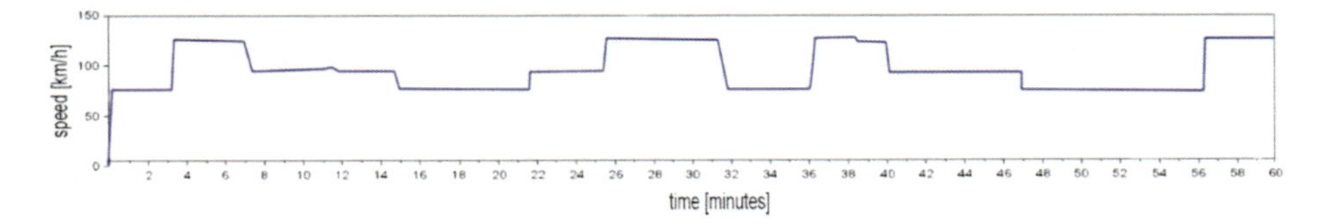

Fig. 2. Leading vehicle changing speed cycles.

4 Data Analyses

4.1 Vehicle Control and Lateral Position

The sleep deprivation has resulted in dangerous lane behaviour characterised by lane departures and weaving for 90% for all subjects, which was verified by the overall increased average deviation from central line (Av.mld, rested: $n = 12$, SD = 0.1, max = 0.6, min = 0.2 mean = 0.4); sleep deprived: $n = 12$, SD = 0.2, max = 0.9, min = 0.2, mean = 0.5) and number of lane departures (NLD, rested: $n = 12$, SD = 49, min = 20, max = 155, mean = 57; sleep deprived: $n = 12$, $s = 497$, min = 19, max = 1808 mean = 143), these values are well-correlated (see Fig. 3). Moreover, increase in number of lane departures is well correlated with increase in lane departure time caused by sleep deprivation (TLS, rested: $n = 12$, SD = 167, min = 13, max = 581, mean = 141; sleep deprived: deprived $n = 12$, SD = 225, min = 49, max = 845, mean = 311) (see Fig. 4).

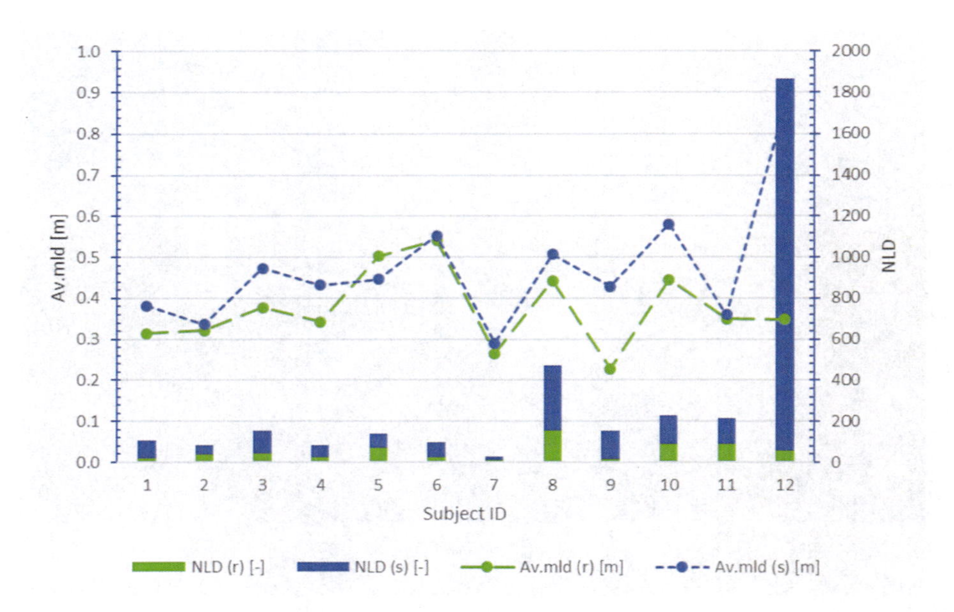

Fig. 3. Correlation of lane departures (LDN) and average mid-line deviation (Av.mdl) in rested (r) and sleep-deprived (s) states.

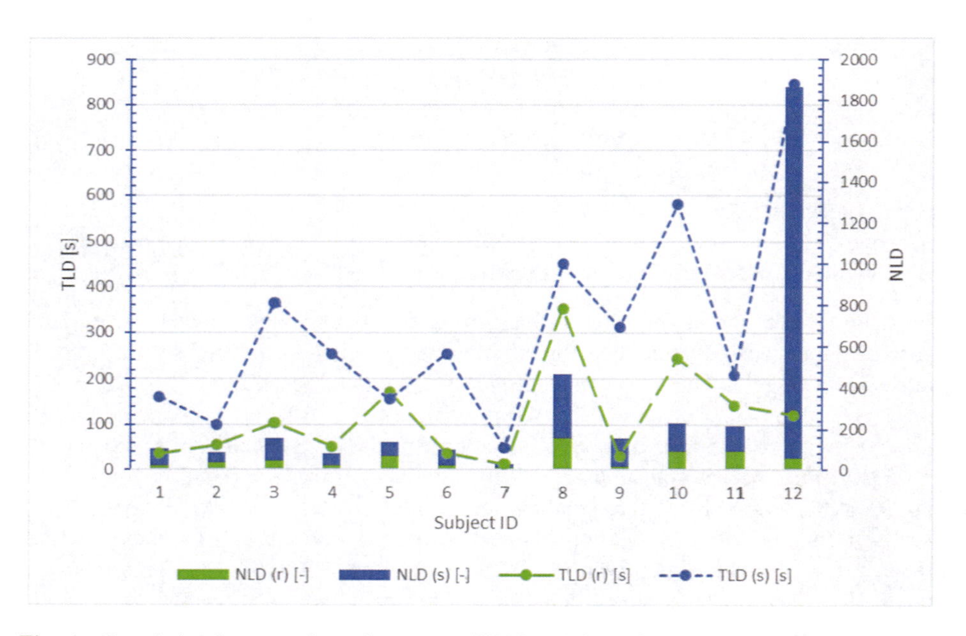

Fig. 4. Correlation between lane departures (LDN and lane departure time (TLD) in rested (r) and sleep-deprived (s) states.

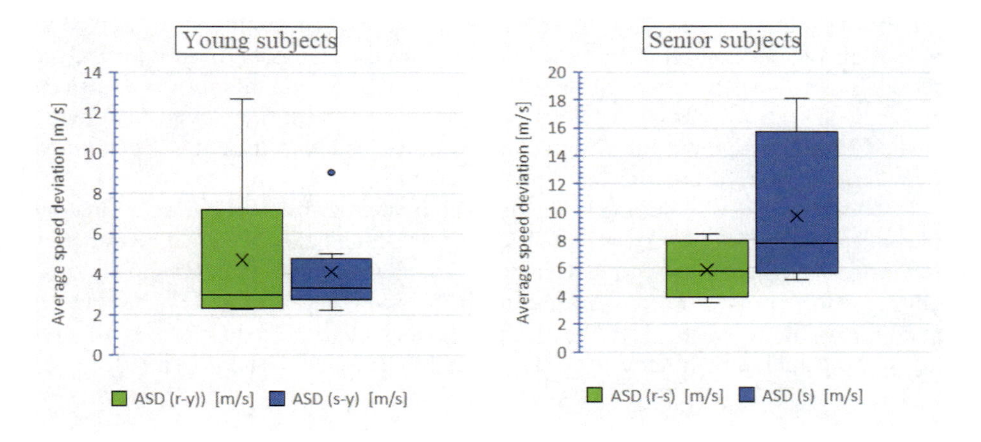

Fig. 5. Deviation of vehicle speed in young (ASD (r/s−y)) and senior (ASD(r/s−s)) subjects.

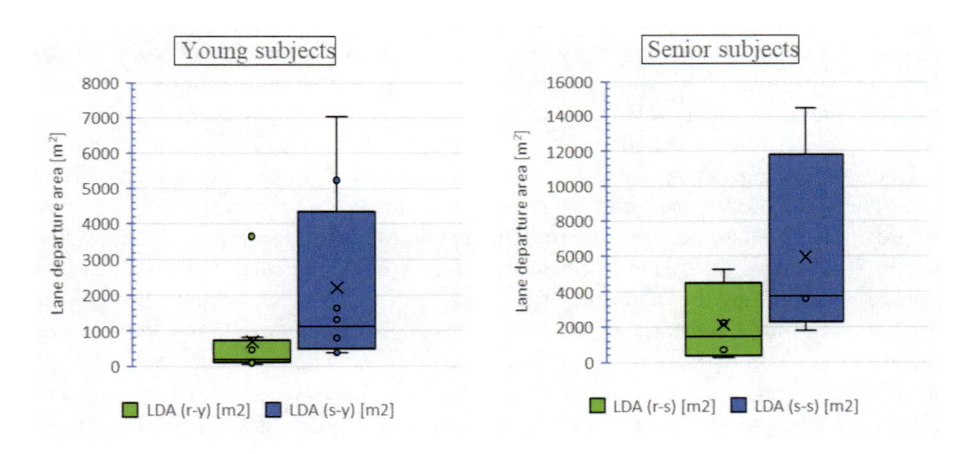

Fig. 6. Lane departure area in young (LDA (r/s−y)) and senior (LDA (r/s−s)) subjects.

No pattern was observed in speed deviation behaviour (ASD), where the differences were individual and might be dependent on personal style or habit of a car following behaviour. However, a regularity was observed when analyses within age/driving experience groups as compared to the baseline state. Thus, the range of speed deviation for younger subjects was significantly narrowed in sleep deprived state, which can be explained by their inexperience and tendency of risky behaviour such as inability to maintain predefined speed and as a result, lead to rear-collision with the leading vehicle in a given task (ASD (y) rested: SD = 3.8, max = 12.66, min = 2.26 mean = 4.7; sleep deprived: SD = 2.2, max = 9.02, min = 2.21, mean = 4.1). On the other hand, a significant increase in range of speed deviation was observed in senior/experienced drivers in sleep deprived state, which on one hand can be explained by increased caution because of the endangered state by maintaining safe distance to the leading vehicle, and on the other hand, is an evidence of inability to fulfil the car following task (ASD(s) rested: SD = 2.1, max = 8.45, min = 5.32 mean = 5.9; sleep deprived: SD = 5.8, max = 18.1, min = 5.18, mean = 9.7); moreover, the range between the groups significantly differed (see Fig. 5). The last statement was confirmed by higher values of NLD(s) for the senior group.

For further analysis of driving behaviour within the two observed groups evaluation of lane departure area (LDA) was performed and was found to be a unified measure for loss of driver vigilance caused by sleep deprivation. The LDA value has shown to be uniformly increasing for both subgroups in a sleep deprived state (see Fig. 6) regardless of differences between the groups' speed and lane behaviour, and is a useful measure to combine both number of lane departure and lane departure time.

5 Summary

The current research was focused on comparison of sleep effects on two age groups: young/novice drivers and senior/experienced drivers. The results imply the challenge of non-uniform behaviour of subjects within each group as well as the individual behaviour. Though separate measures, such as average lane deviation or number of lane departure are an evidence of driver falling asleep, which is consistent with several studies [18, 19], the lane behaviour is still a non-uniform measure and maybe an evidence of an individual driving style or other activities that are not caused by sleep. The speed, however is more relevant to the current research, while its fluctuation has proven to be an evidence of loss of vehicle control due to sleepiness. Besides that, the speed deviation was found to be peculiar for each of the group, where increase in speed deviation may be characteristics of senior/experienced subjects, while the decrease was found typical for younger group. Thus, sleepiness has a more serios impact to the risky behaviour of younger drivers, who are, according to theory of planned behaviour, tend to be more involved in drowsy driving behaviour, which is classified as a risky driving behaviour [20]; moreover, they are more likely to continue driving despite awareness of being sleepy [21].

Acknowledgements. This research was supported by grant SGS16/254/OHK2/3T/16 "Experimental research of driver fatigue by means of observation visual behavior".

References

1. Sunwoo, J.S., Hwangbo, Y., Kim, W.J., Chu, M.K., Yun, C.H., Yang, K.I.: Sleep characteristics associated with drowsy driving. Sleep Med. **40**, 4–10 (2017). https://doi.org/10.1016/j.sleep.2017.08.020
2. Di Milia, L., Smolensky, M.H., Costa, G., Howarth, H.D., Ohayon, M.M., Philip, P.: Demographic factors, fatigue, and driving accidents: an examination of the published literature. Accid. Anal. Prev. **43**(2), 516–532 (2011). https://doi.org/10.1016/j.aap.2009.12.018

3. Watling, C.N., Armstrong, K.A., Radun, I.: Examining signs of driver sleepiness usage of sleepiness countermeasures and the associations with sleepy driving behaviours and individual factors. Accid. Anal. Prev. **85**, 22–29 (2015). https://doi.org/10.1016/j.aap.2015.08.022
4. Sikander, G., Anwar, S.: Driver fatigue detection systems: a review. IEEE Trans. Intell. Transp. Syst. **99**, 1–14 (2018). https://doi.org/10.1109/tits.2018.2868499
5. MacLean, A.W., Davies, D.R., Thiele, K.: The hazards and prevention of driving while sleepy. Sleep Med. Rev. **7**(6), 507–521 (2003)
6. Armstrong, K.A., Smith, S.S., Steinhardt, D.A., Haworth, N.L.: Fatigue crashes happen in urban areas too: characteristics of crashes in low speed urban areas. In: 2008 Australasian Road Safety Research, Policing and Education Conference, Adelaide, South Australia, 10–12 Nov 2008
7. Government of South Australia: Fatigue-related crashes in South Australia (2010). http://www.dpti.sa.gov.au/__data/assets/pdf_file/0010/51022/Fatigue_Fact_Sheet_2010.pdf. Accessed 10 Jan 2019
8. Horne, J.A., Reyner, L.A.: Sleep related vehicle accidents. BMJ **310**(6979), 565–567 (1995)
9. Zhang, G., Yau, K.K., Zhang, X., Li, Y.: Traffic accidents involving fatigue driving and their extent of casualties. Accid. Anal. Prev. **87**, 34–42 (2016). https://doi.org/10.1016/j.aap.2015.10.033
10. Obst, O., Armstrong, K., Smith, S., Banks, T.: Age and gender comparisons of driving while sleepy: behaviours and risk perceptions. Trans. Res. Part F: Traffic Psychol. Behav. **14**(6), 539–542 (2011)
11. Mashko, A.: Subjective methods for assessment of driver drowsiness. Acta Polytech. CTU Proc. **12**, 64–67 (2017)
12. Sandberg, D., Anund, A., Fors, C., Kecklund, G., Karlsson, J.G., Wahde, M., Åkerstedt, T.: The characteristics of sleepiness during real driving at night–a study of driving performance, physiology and subjective experience. Sleep **34**(10), 1317–1325 (2011). https://doi.org/10.5665/SLEEP.1270
13. Mårtensson, H., Keelan, O., Ahlström, C.: Driver sleepiness classification based on physiological data and driving performance from real road driving. IEEE Trans. Intell. Transp. Syst. **20**(2), 421–430 (2019). https://doi.org/10.1109/tits.2018.2814207
14. Lee, M.L., Howard, M.E., Horrey, W.J., Liang, Y., Anderson, C., Shreeve, M.S., O'Brien, C.S., Czeisler, C.A.: High risk of near-crash driving events following night-shift work. Proc. Natl. Acad. Sci. U.S.A. **113**(1), 176–181 (2015)
15. Ingre, M., ÅKerstedt, T., Peters, B., Anund, A., Kecklund, G.: Subjective sleepiness, simulated driving performance and blink duration examining individual differences. J. Sleep Res. **15**, 47–53 (2006)
16. Gharagozlou, F., Mazloumi, A., Saraji, G.N., Nahvi, A., Ashouri, M., Mozaffari, H.: Correlation between driver subjective fatigue and bus lateral position in a driving simulator. Electron. phys. **7**(4), 1196–1204 (2015). https://doi.org/10.14661/2015.1196-1204
17. Li, Z., Chen, L., Peng, J., Wu, Y.: Automatic detection of driver fatigue using driving operation information for transportation safety. Sens. (Basel, Switz.) **17**(6), 1212 (2017). https://doi.org/10.3390/s17061212
18. Classen, S., Shechtman, O., Awadzi, K.D., Joo, Y., Lanford, D.N.: Traffic violations versus driving errors of older adults: informing clinical practice. Am. J. Occup. Ther. **64**(2), 233–241 (2010)
19. Cooper, J.M., Medeiros-Ward, N., Strayer, D.L.: The impact of eye movements and cognitive workload on lateral position variability in driving. Hum. Factors **55**(5), 1001–1014 (2013)

20. Lee, C.J., Geiger-Brown, J., Beck, K.H.: Intentions and willingness to drive while drowsy among university students: an application of an extended theory of planned behavior model. Accid. Anal. Prev. **93**, 113–123 (2016)
21. Jiang, K., Ling, F., Feng, Z., Wang, K., Shao, C.: Why do drivers continue driving while fatigued? An application of the theory of planned behaviour. Transp. Res. Part A: Policy Pract. **98**, 141–149 (2017)

Author Index

A. Varhelyi et al. (Eds.): VISZERO 2018, LNITI, pp. 215–216, 2020.
https://doi.org/10.1007/978-3-030-22375-5

PGSTL